THE

BIRDS OF EUROPE.

BY

JOHN GOULD, F.L.S., &c.

IN FIVE VOLUMES.

VOL. IV.

RASORES. GRALLATORES.

LONDON:

PRINTED BY RICHARD AND JOHN E. TAYLOR, RED LION COURT, FLEET STREET.

PUBLISHED BY THE AUTHOR, 20 BROAD STREET, GOLDEN SQUARE.

1837.

LIST OF PLATES.

VOLUME IV.

Note.—As the arrangement of the Plates during the course of publication was found to be impracticable, the Numbers here given will refer to the Plates when arranged, and the work may be quoted by them.

RASORES.

GRALLATORES.

* Named erroneously in the letter-press American Bittern.

WOOD PIGEON.

Columba palumbus. (Linn.)

Genus COLUMBA.

Gen. Char. *Bill* of mean strength, straight at the base, with the tip or horny point compressed and deflected. Base of the upper mandible covered with a soft, protuberant, cartilaginous substance in which the *nostrils* are lodged towards the middle of the bill, forming a longitudinal cleft. *Feet* with three toes before, entirely divided, and with one hind toe articulated on the heel. *Claws* short, strong and blunt. *Wings* of moderate length and acuminate; the first quill rather shorter than the second, which is the longest.

WOOD PIGEON.

Columba palumbus, *Linn.*

La Colombe ramier.

THE Wood Pigeon, or Ring-dove, is, in Europe, the largest species of the genus to which it belongs, and is sufficiently common over the whole of the European continent to be exceedingly well known, but is considerably more abundant as well as more stationary in the southern parts. It lives principally in woods and forests, and feeds upon all kinds of grain, the leaves of some plants, corn, beech-nuts and acorns. In the British Islands the Wood Pigeon is a constant resident in the large tracts of wooded and inclosed districts, feeding during summer and autumn on the leaves of young clover, green corn, peas, beans, &c., and resorting in flocks during the severer weather of winter to turnip-fields, and to the woods for berries and the harder produce of oaks and other trees.

Early in the spring these birds begin to pair; they make a flat thin nest of small sticks loosely put together, a fir tree in a grove or plantation being a favourite receptacle, on one of the horizontal branches of which the nest is placed, generally twelve or sixteen feet from the ground. The eggs are two in number, oval and white; the young birds are fed from the softened contents of the parent's crop, and two or three pair of young birds, generally a male and a female in each pair, are produced in the season. Ornithologists agree that this species of Pigeon has never been induced to breed in confinement. Montagu says, "We have been at considerable pains to endeavour to domesticate this bird; and though we have tamed them within doors so as to be exceedingly troublesome, yet we never could produce a breed, either by themselves or with the tame Pigeon. Two were bred up together with a male Pigeon, and were so tame as to eat out of the hand, but as they showed no signs of prolificacy in the spring, were suffered to take their liberty in the month of June, by opening the window of the room in which they were confined, thinking the Pigeon might induce them to return to their usual place of abode, either for food or to roost; but they instantly took to their natural habits, and we saw no more of them, although the Pigeon continued to return."

For the information and encouragement of those who may have the inclination as well as the opportunity of making further trials, with the view to endeavour to domesticate so large and valuable a species, we are enabled to state, that a pair of these birds in the dove-house at the Gardens of the Zoological Society in the Regent's Park, built a nest and produced two eggs, but unfortunately during the period of incubation, in which the male assists, the eggs were broken by some of the numerous other birds, most of them of the same genus, with which they were confined.

The head, coverts of the wings and scapulars are of a deep blueish ash colour; the neck in front and the breast vinaceous, beautifully glossed with green and copper colour, changeable in different lights; on each side of the neck is a large patch of glossy white; back and tail ash colour, the latter black at the end; vent and thighs white, tinged with ash colour; the bastard wing almost black, near which a few of the coverts are white, forming a line down to the greater quills, which are dusky, edged with white; beak pale flesh colour, the tip reddish orange; legs and feet red.

Like most of the genus, the Wood Pigeon has great powers of flight. There is little or no distinction in the plumage of the sexes; but the male is the larger bird of the two.

Young birds before their first moult have neither the white space on the sides of the neck, nor the brilliant and glossy appearance of the plumage of adult birds: the whole of their colours also are less pure and decided.

We have figured an old and a young bird of the natural size.

STOCK DOVE.

Columba oenas, (Linn:)

Drawn from life and on Stone by J.& E. Gould.

Printed by C. Hullmandel.

STOCK DOVE.

Columba œnas, *Linn.*

Le Colombe colombin.

ALTHOUGH the Stock Dove closely resembles many of our domestic breeds in plumage and general appearance, it is not now considered the origin of those birds; the European varieties of which are ascribed to another nearly allied species, the *Columba livia*, whose natural habitat is confined to rocks, towers and ruins, especially those adjacent to the sea-coast, whereas that of the present species is restricted to woodlands, building in hollow trees and perching on their branches. Independently of the difference of the localities chosen by each, we find the Stock Dove destitute of that peculiar mark which characterizes most of our domestic varieties, viz. the conspicuous white band on the rump and upper tail-coverts,—a feature equally distinguishing the *Columba livia* from the present bird, in which we find this part of the same lead-coloured blue as the rest of the plumage.

The Stock Dove inhabits the whole of the middle countries of Europe, becoming less common as we approach the northern and southern regions. In England it is found in several midland districts that are well covered with woods, particularly Hertfordshire and the adjoining counties.

The *Columba œnas* is one of the many of its species in which the typical characters both as to colour and form are truly developed : in its manner it closely resembles the Wood Pigeon; it differs, however, in some minor points, viz. the situation of its nest, and its more shy and retired disposition at the time of incubation. Its favourite breeding-place is among parks, beech-woods, and forests abounding with old trees, in the holes of which it frequently breeds, making but little or no nest, as is usual we believe with most of the birds of this genus. It lays two eggs of a pure white. Its food consists of peas, and other leguminous seeds, to which it adds the green tops of turnips and other vegetables.

The sexes offer no differences in plumage, and the young are only to be distinguished by the want of the changeable reflections of green on the sides of the neck, and the general dulness of the plumage.

The head and whole of the upper surface is of a beautiful blueish ash colour; the quill-feathers dark grey; the centre of the wings have two obscure spots of black, the tips of the tail-feathers being of the same colour; the sides of the neck are of a rich changeable green; the breast is tinged with vinous, and the whole of the under parts are of a more delicate grey than the upper; beak dull red; irides and feet dull scarlet. Total length twelve or thirteen inches.

Our Plate represents an adult bird of its natural size.

ROCK DOVE.
Columba livia. *(Linn.)*

Drawn from Nature & on Stone by J & E Gould.   Printed by C Hullmandel.

ROCK DOVE.

Columba livia, *Linn.*

Le Colombe biset.

IT is now almost universally conceded that this small and elegant species of Pigeon is the origin of the numerous domestic varieties which tenant our dovecotes and pigeonlofts. We are aware that the Stock Dove (*Columba œnas*, Linn.) has been reputed as the parent stock from whence they have sprung; but taking the habits and manners of the two species into consideration, we cannot fail to perceive on which side the probability lies, the *Columba œnas* being altogether woodland in its habits and strictly migratory, while the present species is an inhabitant of precipitous rocks, towers, and steeples, never perching on trees, and generally remaining with us throughout the year. To this we may also add, that the white rump which distinguishes the Rock Dove is retained by its descendants in captivity, or if it disappear for a few generations in obedience to the art of the pigeon-fancier, whose skill consists in producing monstrosities in nature, it uniformly returns, or at least shows so strong a tendency so to do that still further crosses are required to prevent the white feathers from predominating. As regards the difference in size and form which our domestic Pigeons exhibit, we have only to observe that they show the same results of the influence of man upon the animals subject to his controul as do the dog, the sheep, and the ox. One thing is certain, as experience has well taught us, that domestication has a decided tendency to produce both an increase of size and a variation of form and colours.

The habitat of the Rock Dove appears to be extended throughout Europe and the greater portion of Africa, particularly its northern regions, everywhere frequenting the rocks which border the ocean, islands, precipices, and ruined buildings adjacent to the coast. Along the shores of the Mediterranean and in the island of Teneriffe it abounds in great multitudes, and in our own islands it is nowhere more plentiful than in the Orkneys and along the coasts of Wales; it is, however, also to be met with wherever abrupt rocks near the sea afford it a congenial asylum.

Like the rest of its genus, it lays two white eggs, on the shelves of the rocks, and is said to breed twice or thrice in the season.

Its food consists of grain and various seeds, to which, according to Montagu, are added the inhabitants of various land shells, particularly that of the *Helix virgata*.

The colouring, which is the same in both male and female, is as follows :

The head, face, and throat deep blueish grey; neck and chest beautiful green and purple, changing in every light; upper and under surface delicate blueish grey, with the exception of the rump, which is white; two distinct bars of black pass across the wings; quills and tail dark grey, the latter tipped with black; bill brown; legs and irides red.

The figure represents the bird of the natural size.

TURTLE DOVE.

Columba Turtur, *(Linn).*

Drawn from Life & on Stone by J & E Gould.   Printed by C. Hullmandel.

TURTLE DOVE.

Columba Turtur, *Linn.*

La Colombe Tourterelle.

AMONG the feathered harbingers of spring, whose voice gladdens our woodlands when "winter is over and gone," this charming bird has ever been celebrated as a universal favourite. The Turtle Dove, as if by common consent, in every age and in every clime, has been considered by the poet as an emblem of serenity and peace, to grace and soften his pictures of rural harmony. No one can listen with indifference to its notes among the budding foliage of the trees, blending with the songs of other birds, and harmonizing with the genial influence of reviving Nature.

The Turtle Dove appears among us in April; but, like other birds that arrive about the same period, its appearance is influenced by the congeniality and mildness of the season. On arriving, it takes up its abode in the thickly wooded districts of our island, more especially the midland and southern counties, commencing the process of incubation as soon as the foliage becomes sufficiently dense to afford it shelter, selecting with indifference any tree, but more frequently the fir and such others as have their stems covered with ivy, and thus afford a secure resting-place for their rude flat nest. Upon this frame-work, composed of a few straight sticks most inartificially crossed, and interwoven with little care or skill, the female deposits two eggs of the purest white, which may be seen through the apertures of the nest, so slightly and so rudely is it built.

The Turtles pair, as do the Doves in general. The male and female sit by turns, alternately relieving each other, dividing the task of incubation and mutually providing for the wants of their unfledged progeny. The young the first autumn have only the indication of white on the neck, and the feathers of their plumage emarginated with distinct and lighter brown. They as well as their parents depart, in September and October, to the opposite shores of the European Continent, whence they proceed southward to more congenial climes. The Turtle Dove, however, is abundant over Continental Europe, extending far northward, but is not found, we believe, within the region of the arctic circle. Its general habits appear to be migratory; and if we may hazard a conjecture, we should consider that the northern and particularly the more woody portions of the coast of Africa form its winter domicile, as we have reason to know is the case with so many of our summer visitants. Its food consists of grain and vegetables, in search of which it frequents fields of corn for pease and other vegetable seeds.

The sexes differ in so trifling a degree, that the description of one will serve for both. The head and neck are varied with ash-colour, becoming richer and brighter on the breast; the sides of the neck are distinguished by a patch of black feathers uniformly tipped with white, so arranged as to produce a series of alternate lines of black and white; the back dark brown; the wing-coverts reddish brown, each feather having a large dark central mark; tips of the shoulders lead-colour; quill-feathers brown; the lower part of the belly and tail-coverts white; tail rounded; the two middle tail-feathers brown; the rest tipped with white; the external feather on each side having its external edge also white; irides and feet red. Length eleven inches.

Our Plate represents an adult male.

COMMON PHEASANT.
Phasianus Colchicus. (Linn.)

Genus PHASIANUS, *Linn.*

GEN. CHAR. *Bill* of mean length, strong; upper mandible convex, naked at the base, and with the tip bent downwards. *Nostrils* basal, lateral, covered with a cartilaginous scale; cheeks and region of the eyes destitute of feathers, and covered with a verrucose red skin. *Wings* short, the first quills equally narrowed towards their tips, the fourth and fifth the longest. *Tail* long, regularly wedge-shaped, and composed of eighteen feathers. *Feet* having the three anterior toes united by a membrane as far as the first joint, and the hind toe articulated upon the tarsus, which in the male birds is furnished with a horny, cone-shaped, sharp spur.

COMMON PHEASANT.

Phasianus Colchicus, *Linn.*

Le Faisan vulgaire.

THIS bird has been so long naturalized that it may now be said to claim a place in the European Fauna; it would, however, appear that Europe is not its aboriginal habitat, and that there is every reason to believe that it was introduced at a very early period from the western confines of Asia; and history assigns to Jason the honour of having brought it from the banks of the Phasis on his celebrated expedition; and from whence the various modifications of the word are derived, viz. Phasianus in Latin, Pheasant in our own language, Faisan in French, Faisiano in Italian, &c. The ancient Colchis is the Mingrelia of the present day, and here it is said to be still found wild and unequalled in beauty.

All the details connected with the habits and manners of this species are so well known to every one that we need do little more than refer our readers to the minute and accurate descriptions published by Montagu and Mr. Selby, of the changes of plumage, diseases to which it is subjected, &c.

The nest is very inartificial, and is placed on the ground in long grass or thick underwood, and not unfrequently in fields of clover: the eggs are of a clear dull green, and from ten to fourteen in number. The young, which are hatched during the months of June and July, continue with the females until they begin to moult and assume the adult plumage, which commences about the beginning of September, and is completed by the middle of October.

The food of the adults consists of grain and seeds in winter, of roots and insects in spring and summer; but the young are fed exclusively upon the latter. "I have observed," says Mr. Selby, "that the root of the Bulbous Crowfoot (*Ranunculus bulbosus*), a common but acrid meadow plant, is particularly sought after by this bird, and forms a great portion of its food during the months of May and June. The root of the garden tulip is also an article of diet, which it omits no opportunity of obtaining, and which by means of its bill and feet, it is almost certain to reach, however deep it may be buried."

The male has the cheeks naked and of the brightest scarlet, minutely speckled with black; the crown of the head bronzed green; on each side of the occiput a tuft of dark golden green feathers, capable of being erected at pleasure, and very conspicuous in the pairing-season; upper part of the neck dark green, glossed with purple and violet blue; lower part of the neck, breast, and flanks deep reddish orange, showing in some positions beautiful reflections of light purple; each feather bordered and terminated with pansy purple; centre of the belly and thighs blackish brown; centre of the back and scapular feathers black or brownish black, surrounded with a yellowish white band and bordered with deep reddish orange; lower part of the back and upper tail-coverts green, intermingled with brownish orange and purplish red; tail-feathers brown, crossed by bands of black, and fringed with reddish brown; bill pale yellow; irides pale brownish orange; legs and toes greyish black.

The female has the cheeks covered with small closely set feathers, and the whole of the plumage yellowish brown, mingled with different shades of grey, brown, and black.

We have figured an adult male and female one third less than the natural size.

CAPERCAILZIE OR COCK OF THE WOOD.

Genus TETRAO, *Linn.*

GEN. CHAR. *Bill* short, strong; upper mandible convex, and arched from the base to the tip. *Nostrils* basal, lateral, partly closed by an arched scale, and hidden from view by small closely set feathers. *Eyebrows* naked, and adorned with a red papillose and fringed skin. *Wings* short; first quill much inferior in length to the second, which is shorter than the third and fourth. *Tail* of sixteen feathers. *Feet* with three toes before, united as far as the first joint, and one behind, short; the edges of all pectinated. *Tarsi* feathered to the toes.

CAPERCAILZIE, OR COCK OF THE WOOD.

Tetrao Urogallus, *Linn.*

Le Tetras Auerhan.

THIS noble bird, the largest of the Grouse tribe, was once common in the forests of Scotland, and, in all probability, in the northern parts of our island; for centuries, however, it has been gradually diminishing in number until at length it has become completely extinct there. Mr. Selby informs us that "the last individual of this species in Scotland was killed about forty years ago, near Inverness; previous to which date the breed had become extinct in Ireland."

Those who wish to see this bird in a state of nature must visit the extensive forests which cover the greater portion of Norway and Sweden, from whence our markets are annually supplied with this celebrated game. Although these countries may be considered as its great depôt, it is also to be found in all the wild and mountainous districts of Europe wherever extensive pine-woods afford it food and shelter.

In habits and disposition the Capercailzie and the Black Cock (*Tetrao Tetrix*), differ considerably from the genuine Grouse; they are, in fact, more essentially arboreal, and their feet being furnished with horny papilli enable them to rest on the smooth and slippery branches of the pine and other alpine trees. As the breeding-season approaches, the male becomes greatly excited, and perched on some large branch, in a dense part of the forest, invites his mate by often-repeated and loud calls, which very frequently leads to his own destruction by apprising the sportsman of the immediate locality in which he is seated: hence in the months of April and May the London markets are frequently stocked with fine males in their highest state of plumage, and with a favourable wind for the lobster-boats, which are the general means of conveyance, arrive in good order.

The male far exceeds his mate in size and in the beauty and elegance of its plumage.

The Capercailzie is decidedly polygamous in its habits, and lives separate from the females except in the breeding-season. The female rears her young in independent seclusion: the nest is placed amidst brakes and dense underwood; the eggs are from eight to sixteen in number, of a yellowish white spotted with darker yellow.

The young of both sexes during the first autumn resemble the female; but as spring approaches, the males assume their sexual characteristics of plumage.

The food of the Capercailzie consists of alpine berries, tops of fir, snails, &c.

The male has the head, neck, whole of the upper surface, flanks and under tail-coverts dark ashy grey, with innumerable small irregular markings of black; forehead and throat dusky black, the feathers of the latter being long and pendulous; breast fine dark glossy green; wings fine chestnut brown, with innumerable small irregular markings of black; secondaries slightly tipped with white; quills dull brown on their inner webs, paler on the outer; under surface black with spots of white, which are most numerous about the thighs and vent; the two outer rows of the upper tail-coverts are tipped with white, are considerably longer than those in the middle, and, gradually lengthening, reach nearly to the end of the tail, which is rounded and of a black colour with a few irregular spots of white on the sides; bill yellowish white; irides hazel; over the eye a bare red skin; legs covered with brown hair like feathers.

The female has the head, neck, and upper part of the back barred with reddish brown, grey and black; wings and lower part of the back dark brown, each feather edged and tipped with irregular markings of reddish brown; secondaries tipped with white; quills as in the male but lighter; throat pale reddish brown; breast rich reddish brown; under surface pale reddish brown, each feather being barred near its extremity with dark brown and tipped with greyish white, which colour predominates on the under tail-coverts; tail rich rufous brown, numerously barred with very dark brown; bill greenish horn colour; legs pale brown.

We have figured a male and female about two thirds of the natural size.

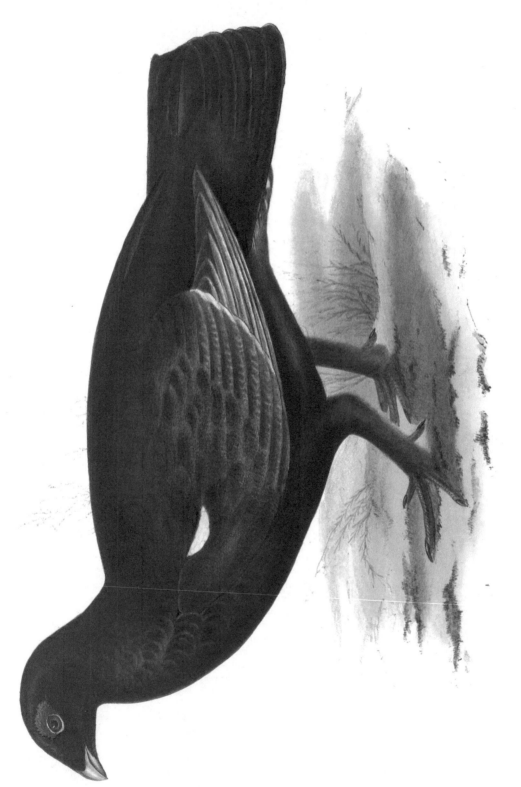

HYBRID GROUSE.
Tetrao hybridus. (Sparrm.)

HYBRID GROUSE?

Tetrao hybridus, *Sparrm.*

—— medius, *Meyer.*

Le Tetras Rakhelhan.

WE believe it is now the prevailing opinion among naturalists, that the bird figured in the accompanying Plate (the *Tetrao medius* of Meyer and Temminck,) is merely a hybrid between the Capercailzie and the Black Grouse, and as we ourselves are inclined to entertain the same opinion, our figure will therefore be of interest as an illustration of the singular appearance presented by the mixture of the two species. While in this country, Professor Nilsson of Stockholm, a competent judge, and in whose opinion we should consequently have no hesitation to confide, informed us that no doubt existed in his mind as to the hybridization of the two birds in question, and this we believe is also the opinion of the resident inhabitants of the countries where they are found. On a subject respecting which so much doubt exists, it will perhaps be preferable to consider it as a hybrid, until the discovery of a female or more information is acquired relative to its history. The countries in which this bird is found are of course only those inhabited by the two species in question, consequently Norway and Sweden are the places where it is most generally met with. Few springs pass without examples of this bird being sent to London in company with Capercailzies, Black Grouse, &c. Of the numerous individuals we have had opportunities of examining little difference was found to exist; neither, we believe, did the examples inspected by Mr. Yarrell differ from each other in their internal structure.

Head, the whole of the neck, and breast dark purple; the feathers of the head, and back of the neck very minutely freckled with whitish; the remainder of the upper surface dark brown, minutely freckled with lighter brown; the scapularies and secondaries terminated with whitish; quills dull brown margined with greyish white; on the shoulder a small patch of white; under surface dark brown minutely freckled with lighter brown; vent white; under tail-coverts black, largely tipped with white; tail black, some of the centre feathers slightly tipped with greyish white; feathers of the legs mingled with greyish white and brown; bill horn colour; feet black.

The figure is of the natural size.

BLACK GROUSE.
Tetrao Tetrix, (Linn.)

BLACK GROUSE.

Tetrao Tetrix, *Linn.*

Le Tétras Birkhan.

THE European portion of the globe may be truly considered the exclusive habitat of this noble bird; if however any exception is to be made to this rule it will probably be in favour of Siberia, but of its existence there we have no certain information.   On the Continent it occurs commonly in Russia, Norway, Sweden, Germany, France, and some parts of Holland.   From Norway and Sweden great numbers are annually brought to the London markets, and, together with the Capercailzie and Willow Ptarmigan, forms no trifling article of commerce.   At the same time it is far from being uncommon in many parts of England, and in Scotland it is very abundant.   Mr. Selby, who is at once a naturalist and a sportsman, has so well described the manners and habits of the Black Grouse, that we trust we shall be pardoned for transcribing the observations of this gentleman, whose splendid work on British Ornithology is too well known to need any eulogium from our pen.

" The present species is now confined, in the southern parts of England, to a few of the wildest uncultivated tracts, such as the New Forest in Hampshire, Dartmoor, and Sedgemoor in Devonshire and the heaths of Somersetshire.   It is also sparingly met with in Staffordshire and in parts of North Wales, where it is under strict preservation.   In Northumberland it is very abundant, and has been rapidly increasing for some years past, which may be partly attributed to the numerous plantations that within that period have acquired considerable growth in the higher parts of the county, as supplying it both with food and protection.   It abounds throughout the Highlands of Scotland, and is also found in some of the Hebrides.   The bases of the hills in heathy and mountainous districts which are covered with a natural growth of birch, alder and willow, and intersected by morasses, clothed with long and coarse herbage, as well as the deep and wooded glens so frequently occurring in such extensive wastes, are the situations best suited to the habits of these birds, and most favourable to their increase.   During the months of autumn and winter the males associate and live in flocks, but separate in March or April; and, being polygamous, each individual chooses some particular station, from whence he drives all intruders, and for the possession of which, when they are numerous, desperate conflicts often take place.   At this station he continues early every morning, and in the evening during the pairing-season, repeating his call of invitation to the other sex, and displaying a variety of attitudes, not unlike those of a Turkey Cock, accompanied by a crowing note, and by another similar to the noise made by the whetting of a scythe.   At this season his plumage exhibits the richest glosses, and the red skin of his eyebrows assumes a superior intensity of colour.   With the cause that urged their temporary separation, their animosity ceases, and the male birds again associate, and live harmoniously together.

" The female deposits her eggs in May; they are from six to ten in number, of a yellowish grey colour blotched with reddish brown.   The nest is of most artless construction, being composed of a few dried stems of grass placed on the ground, under the shelter of a tall tuft or low bush, and generally in marshy spots, where long and coarse grasses abound.   The young of both sexes at first resemble each other, and their plumage is that of the hen, with whom they continue till the autumn moult takes place; at this time the males acquire the garb of the adult bird, and quitting their female parent, join the societies of their own sex.   The food of the Black Grouse during the summer, chiefly consists of the seeds of some species of *Juncus*, the tender shoots of heath, and insects.   In autumn, the crawberry or crawcrook (*Empetrum nigrum*), the cranberry (*Vaccinium oxycoccos*), the wortleberry (*Vaccinium vitis-idæa*), and the trailing arbutus (*Arbutus uva-ursi*), afford it a plentiful subsistence.   In winter and during severe and snowy weather, it eats the tops and buds of the birch and elder, as well as the embryo shoots of the fir tribe, which it is well enabled to obtain, as it is capable of perching upon trees without any difficulty.   At this season of the year, in situations where arable land is interspersed with the wild tracts it inhabits, descending into the stubble-grounds, it feeds upon grain.

" In the adult state the Black Grouse displays great shyness of character, and after the autumnal moult, is not easily approached within gunshot.   Frequent attempts have been made to domesticate this bird, but without success; and through all the trials that have taken place, it has never been known to breed in confinement."

The flesh of this species is in such general request as a delicacy for the table, that we need make no remark on its excellence as an article of food.

The male has the head, neck, breast, back, and rump black with purple and blue reflections; abdomen, wing-coverts, and tail deep black; secondaries tipped with white, forming with the adjoining coverts a band across the wing; under tail-coverts pure white; feathers of the legs blackish grey; bill black; feet brown.

The female has the head, neck, all the upper surface and tail orange brown blotched and rayed with black; breast and under surface pale brown barred with black and brown; under tail-coverts white, rayed with black; bill and feet as in the male.

We have figured an adult male and female, rather less than the natural size.

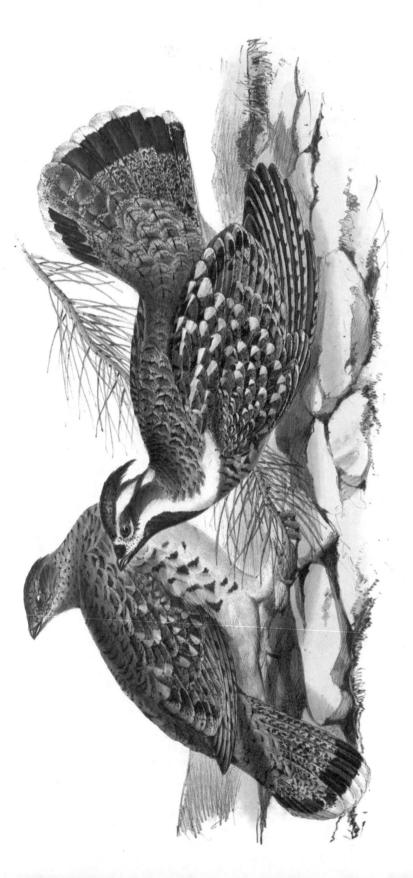

HAZEL GROUS *OR* GELINOTTE.
Bonasia Europea.

Genus BONASIA.

GEN. CHAR. *Head* crested; *tarsi* and *toes* unplumed; and a tuft of feathers projecting from each side of the neck; in other respects as in the genus *Tetrao*.

HAZEL GROUSE, OR GELINOTTE.

Bonasia Europæa.

Le Tetras Gélinotte.

IN the days of Linnæus and the older writers every species of Grouse was comprised in a single genus (*Tetrao*); subsequent research, however, having added many other species, and even new forms, modern naturalists have been induced to subdivide this interesting family into smaller groups, each distinguished by characters peculiar to itself: it has, moreover, been considered expedient to apply to each of these minor subdivisions a generic title, retaining the old name *Tetrao* for the most typical Grouse; the Ptarmigan necessarily forms another group, and the bird under consideration a third. The half-plumed tarsi, the crested head, and the tuft of feathers on each side of the neck are features peculiar to the genus *Bonasia*: in the European Hazel Grouse this latter character is but slightly indicated, but is exhibited to a greater extent in a species from America. The *Bonasia Europæa* is the only species yet discovered in the Old World, but it has its representative in the New, in the well-known Canada Grouse, *Bonasia umbellus*, and others. These slight differences in structure are, as might be supposed, always accompanied by some difference of habits. The feet of the Ptarmigan are as ill adapted for perching on trees as the pectinated toes of the Capercailzie are expressly fitted for that purpose. Although the Hazel Grouse does not equal the Ptarmigan in flight, its powers in this respect are far from being inconsiderable. They frequently perch on trees, and love to dwell in wooded plains skirting hilly and mountainous districts; they feed on alpine fruits and berries, to which are added the tops of heath, fir, juniper, and other tender shoots. They fly in packs or companies, and are not so shy or distrustful as most other members of this family; when disturbed they perch on trees, and are then easily approached and shot.

The Hazel Grouse is dispersed over the continent of Europe from north to south, inhabiting nearly all the elevated ridges and natural boundaries of the different countries. Dr. Latham states that they are so abundant on a small island in the gulf of Genoa that the name of Gelinotte Island has been given to it. It also inhabits France, Germany, Sweden, Norway, and Russia, thus extending itself from the sultry regions of Italy to the limits of the arctic circle. No instance is on record of its having been discovered in England, and so exclusively does it appear to be confined to the European continent, that we have never observed it in collections from other countries.

The eggs of the Hazel Grouse are from ten to twelve in number, of a rusty red, thickly spotted with a darker colour, and are deposited on the ground at the foot of a fern or hazel-stem.

The male may be distinguished from the female by the red naked skin behind the eye, the black mark on the throat, and by the bright and more contrasted markings of the plumage. The young during the first autumn are without the black throat, in which state the colouring of the plumage nearly resembles that of the female.

The Hazel Grouse is held in high esteem for the table, its flesh being both delicate and of good flavour; for which purpose thousands are yearly captured; and it not unfrequently happens that small packages of them in good preservation arrive from Norway and Sweden, in the markets of our Metropolis.

The male has the top of the head, crest, and upper surface varied with markings of reddish brown, black, and grey, disposed in zigzag lines across every feather; the scapularies and secondaries having a large spatulate mark of white running down the stem of each feather; the primaries are brown on their inner webs and varied with buff and brown on the outer; the feathers of the tail, with the exception of the two middle ones, which are grey finely freckled with black, are strongly banded with black near their tips, which are grey; the throat is black, encircled with an obscure band of white, which extends to the shoulders; naked skin over the eye scarlet, bounded above by a patch of white; the feathers of the chest and flanks are black and red with a white tip; those of the breast and belly and the tail-coverts are white, each having a black centre; bill, feet, and eyes brown.

The female differs from the male in not having the black throat and the red naked skin over the eye, and in being less brilliant in all her markings.

The Plate represents a male and female of the natural size.

RED GROUSE.
Lagopus Scoticus (Leach)

RED GROUSE.

Lagopus Scoticus, *Lath.*

Le Tetras rouge.

THE Red Grouse, so renowned for the delicious flavour of its flesh, and so highly prized by the sportsman for the amusement it affords him while in pursuit of his favourite occupation, is so exclusively a native of the British Isles that it has never been discovered either on the adjacent continent or upon that of America. It is not a little surprising that a bird so widely spread over all the heathy districts of our islands, especially those of Scotland, Yorkshire, Westmoreland, Cumberland, Derbyshire, Wales, and many parts of Ireland, should be so strictly confined in its habitat as to be unknown in any other part of the globe, more especially as the Black Cock and Ptarmigan, neither of which possesses greater powers of flight, are dispersed over a large portion of its more northern latitudes. Wide and open moors and heaths, particularly such as are characterized by swelling hills and undulations, are the situations to which the Red Grouse gives preference. It pairs and commences the task of incubation very early in the spring, the female laying in March or April. The young keep in the company of their parents until the autumn, when the various broods assemble together and form large flocks, called *packs* by the sportsman, which continue associated till the spring, when, in obedience to the great law of nature, each selects its mate, and they then disperse over the moorlands to commence the work of reproduction.

Its food consists of the tender tops of the heath, and the fruit of the bilberry, cranberry, and various plants of the genus *Arbutus*; they also readily eat oats and other grain, hence those farmers whose lands adjoin heathy districts often suffer very considerably from their visits. Their flight is rapid, and is often sustained for a considerable distance, particularly after being harassed during the early part of the shooting season, which commences with these fine birds on the first of August.

They construct little or no nest, the eggs, which are from eight to twelve in number, of a reddish white blotched all over with dark brown, being deposited in a shallow cavity lined with a few loose grasses, generally placed in a tuft of heath.

The sexes may be distinguished by the male being darker in colour and by his having the red naked skin over the eye larger and of a more intense colour than in the female.

The whole of the plumage is of a rich chestnut brown marked with fine undulating bars of black, and often irregularly blotched with white; the belly and vent-feathers white; tail black, with the exception of the four middle feathers, which are ash brown barred with black; tarsi and toes clothed with greyish white downy feathers; naked skin over the eye red; irides hazel; bill and nails black.

The Plate represents a male and female of the natural size.

COMMON PTARMIGAN.

Lagopus mutus. *(Leach)*

Drawn from Nature & on Stone by J. & E. Gould.

Printed by C. Hullmandel.

COMMON PTARMIGAN.

Lagopus mutus, *Leach.*

Le Tetras Ptarmigan.

THE great care which Nature takes of her subjects is beautifully exemplified in the mountain Ptarmigans, whose habits and manners lead them to dwell in situations where they experience the greatest extremes of temperature, but against the effects of which they are most amply provided, not only by the assumption of a thick under-covering on the approach of winter, but by a total change in the colour of the plumage, which assimilates to the surface around them, and doubly tends to their safety and preservation, and renders them much less conspicuous to their enemies.

As might be supposed, the mountain Ptarmigans are less wary and shy in their disposition than the other Grouse, doubtless in consequence of being less disturbed by man, against whom the elevated regions they inhabit present an obstacle of too formidable a nature to be often encountered. The common Ptarmigan appears to enjoy an extensive range throughout the whole of the alpine districts of the middle of Europe, as well as in the northern part of the American continent; it is also found, but in less abundance, in Norway, Sweden, and Russia, countries in which the *Lagopus saliceti* is more especially diffused. In the British Islands it is found in all the mountain districts of Scotland, and it is reported to have inhabited Wales at a former period.

Our Plate will convey better than any description we can give, the great difference which exists between the plumage of summer and winter; a change, we may observe, which takes place by the process of a gradual moult.

The Ptarmigan's food consists, in summer, of the berries of alpine plants, and the young shoots of heath in the winter: when the mountains are covered with snow, it burrows beneath it in search of food, as well as for protection against the severities of the season. It incubates early in spring: the eggs, which are from twelve to fifteen in number, have a white ground colour mottled all over with reddish purple brown, and are placed, without any nest, on the bare ground. In the colour of their plumage, the young resemble the female in summer, and gradually change to white with the approach of winter.

The sexes are only to be distinguished by the somewhat larger size of the male, and the more intense black streak between the bill and the eyes.

In summer, the whole of the upper surface is minutely barred with black and deep ochreous yellow; the feathers of the breast and flanks are also of the same colour; the outer feathers of the tail are black; the under surface is greyish white; the primaries white; the shafts black, and these feathers are only moulted once during the year.

In winter, the whole of the plumage is pure white, except the outer tail-feathers, and the spot between the bill and the eyes, which are black.

The Plate represents an adult in the summer plumage, and one in the snowy livery of winter.

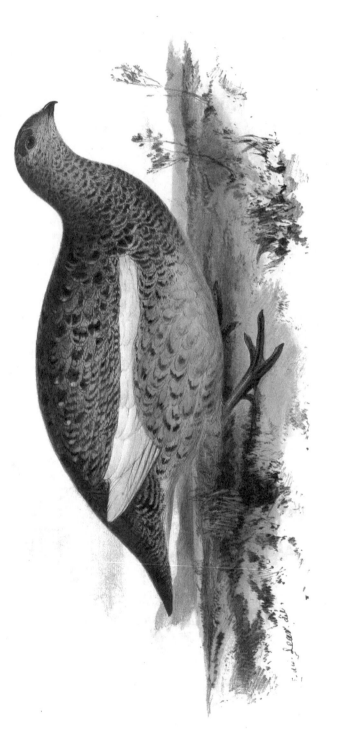

ROCK PTARMIGAN.

Lagopus rupestris, (Leach).

E. Lear del. et lith.

Printed by C. Hullmandel.

ROCK PTARMIGAN.

Lagopus rupestris, *Leach.*

WE are indebted to the kindness of the Earl of Derby for a drawing of a bird in His Lordship's collection which has been considered by some as identical with the *Lagopus rupestris* of North America; while on the other hand several of the best ornithologists have after a minute examination been induced to consider that this, and the other birds killed in Scotland in the orange-coloured dress here represented, are nothing more than the females of the Common Ptarmigan (*Lagopus mutus*) in the plumage of spring and summer, during which period of the year, as abundance of examples testify, both sexes of that bird have the plumage strongly marked with rufous and brown; the male, however, has the colouring much more grey, the rufous markings being principally on the neck and breast, where it sometimes approaches to black. Should this prove to be the case, our figure will not be destitute of interest, as it will exhibit the female Ptarmigan in a state of plumage not usually seen; but until this matter is finally cleared up we have thought it best to figure the bird under the specific title of *rupestris*. The drawing above mentioned, and the figure recently published by T. C. Eyton, Esq., in his continuation of Bewick's British Birds, were both taken from the same specimen. In conclusion, we ourselves question whether any of the American Grouse can be referred with certainty to those of the European continent, for on comparison they always present differences which, although minute, are nevertheless constant.

The whole of the wings are white; the tail black; all the remainder of the plumage rich orange brown, numerously marked and barred with blackish brown, but least so on the belly and flanks; bill and claws black.

Our figure is of the natural size.

WILLOW PTARMIGAN.
Lagopus Saliceti. (Swains.)

WILLOW PTARMIGAN.

Lagopus Saliceti, *Swains.*

Le Tétras des Saules.

THE characteristic features which distinguish the Willow Ptarmigan from the *Lagopus mutus* consist in its superior size, in the rufous colouring of its summer plumage, and the total absence at all times of the black mark between the bill and the eye. Although not an inhabitant of the British Islands, the Willow Ptarmigan is perhaps more widely dispersed than any other species, being abundant over the whole of the arctic circle. It is the most plentiful kind of Grouse found in Norway, Sweden, and Lapland; and extends its range over the whole of Russia and Siberia. These elevated and dreary regions afford it situations most congenial to its habits and mode of life, in consequence of which its numbers are much diminished as it approaches more temperate climes; hence in the midland countries of Europe it is much less frequently seen, and south of these latitudes it is never found.

In its habits, manners, and general economy it strictly resembles the common species, feeding during summer on the tender shoots and buds of heath, together with berries of alpine plants: in winter, when the face of the country is covered with snow, it burrows beneath the surface, and feeds on the scanty herbage, the buds of the dwarf willow, and whatever green vegetable food it can obtain.

Its nest is placed on the ground among tufts of herbage and brushwood : the eggs are from six to ten in number, larger than those of the Common Ptarmigan, but much resembling them in colour.

There is perhaps a greater contrast between the summer and winter plumage of this bird than in any other of the Ptarmigans. The purity of the white in winter being contrasted by the rich colouring of summer, which in some individuals we have seen is of a pure uniform chestnut, with scarcely any trace of the zigzag bars of black.

Our Plate exhibits two birds, one in the pure white livery of winter, the other in an intermediate stage, namely, that of spring, the white having to a great extent given place, by a partial moult, to the coloured feathers of summer.

The sexes offer little difference at either season of the year. The first plumage of the young is coloured, which at the autumn moult is exchanged for white.

In summer the head, neck, back, scapularies, middle tail-feathers and coverts are of a pure chestnut more or less deep, and more or less blotched, with zigzag lines of black; breast, vent, centre of the wings, and quill-feathers pure white; lateral tail-feathers black; beak and nails horn colour; irides greyish white.

In winter the whole of the plumage is pure white.

The Plate represents two adult birds, in different stages of plumage, of the natural size.

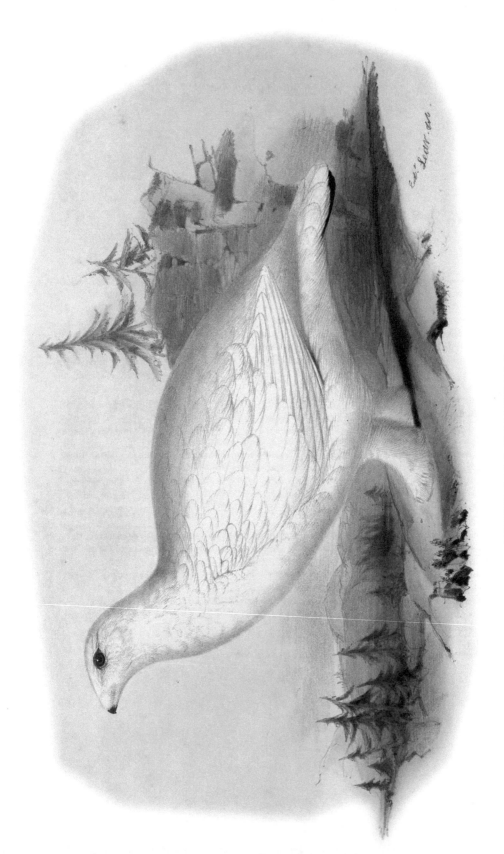

SHORT-TOED PTARMIGAN.
Lagopus brachydactylus.
Tetrao _____. (Temm.)

Ed.ʳ Lear. del

SHORT-TOED PTARMIGAN.

Lagopus brachydactylus.

Tetrao brachydactylus, *Temm.*

Le Tetras à doigts courtes.

WE are indebted to the kindness of M. Temminck for the loan of his specimen of this fine species, which he informs us he received from the Baron de Feldegg of Frankfort, who obtained it from the northernmost part of Russia. It is a remarkable and well-defined species, differing from *Lagopus Saliceti* in having the nostrils thickly covered and the bill almost hidden by feathers, in its shorter and thickly feathered tarsi, and in having the shafts of the primaries and the nails of the toes of a pure white. We regret that we are unable to give any account of its habits and manners, in which it doubtless closely resembles the other members of its genus; or of the localities it frequents, further than that it appears to be an inhabitant of the extreme northern regions of the Old World. It is exceedingly interesting to the ornithologist, from the circumstance of its forming another species of that beautiful group the Ptarmigans.

If we were more fully acquainted with its history, we should doubtless find that it is subjected to a change of plumage similar in every respect to its congeners, being of a rich chestnut brown in summer, and of a spotless white in winter. M. Temminck's specimen, which we believe to be unique, has the whole of the plumage of a pure white, with the exception of the tail-feathers, which are black; the bare skin over the eye scarlet; and the bill black.

Our figure is rather less than the natural size.

SAND GROUSE.

Pterodes arenarius. (Temm.).

Genus PTEROCLES, *Temm.*

GEN. CHAR. *Beak* moderate, compressed, sometimes slender ; the upper *mandible* straight, and curved towards the point. *Nostrils* basal, partly closed by a membrane, and covered by the feathers of the forehead. *Feet* furnished with short toes, the hind toe being very small, and articulated high on the tarsus ; three toes before united together as far as the first articulation and edged by a membrane ; the front of the tarsus covered with short feathers, the posterior aspect being naked. *Nails* very short, that of the hind toe sharp, the rest obtuse. *Tail* conical ; the two middle feathers occasionally prolonged into filaments. *Wings* long and pointed ; the first *quill-feather* longest.

SAND-GROUSE.

Pterocles arenarius, *Temm.*

Le Ganga unibande.

THE birds which compose the genus *Pterocles* have been separated by M. Temminck from the genus *Tetrao*, under which they had been previously included ; and, as they now stand, form a well-defined genus, the species of which inhabits the dry and sandy deserts of the hotter portions of the globe. The number, however, is far from being considerable ; two only have, we believe, as yet been ascertained to be natives of Europe. The present species is found in many of the provinces of Spain, particularly Granada and Andalusia ; it is also found in Sicily and in the deserts of the North of Africa, as well as in Asia, whence we have received it from the Himalaya Mountains. It does not appear to extend itself further northward in Europe than the places above noticed.

The *Pterocles arenarius*, as far as our experience goes, is the largest of its genus, and belongs moreover to that section of it which wants the elongated filiform tail-feathers, so characteristic of the other European species. In habits, manners, and places of nidification, it closely resembles the *Pterocles setarius*,—circumstances which we have detailed at length in our description of that beautiful bird.

The male and female differ considerably in their plumage,—a circumstance in a greater or less degree characteristic of the species of this genus.

In the male, the top of the head, the occiput and breast are of a delicate grey colour ; the back and wing-coverts light rufous ; each feather being irregularly blotched with greyish black and tipped with tawny yellow ; the quill-feathers dark grey ; the throat and sides of the cheeks rufous, beneath which a large triangular black mark surmounts the delicate grey of the breast, and across this again extends a black band passing from one shoulder to the other. The whole of the under surface is black, with the exception of the extremities of the tail-coverts, which are white ; the tail tawny yellow with grey bars, and terminating gradually with the same colour. Total length twelve or thirteen inches.

In the female, the whole of the upper surface is of a tawny yellow, thickly covered with irregular zigzag and barbed markings of black ; the throat merely affords an indication of the black mark which distinguishes the male ; the breast is of the same colour as the upper surface, spotted with black and crossed from shoulder to shoulder with a narrow band of black, beneath which, and the under surface which is black, intervenes a space of about an inch broad of a delicate fawn colour ; the under tail-coverts partly white as in the male ; feet greenish olive.

We have figured a male and female in their adult plumage, rather less than their natural size.

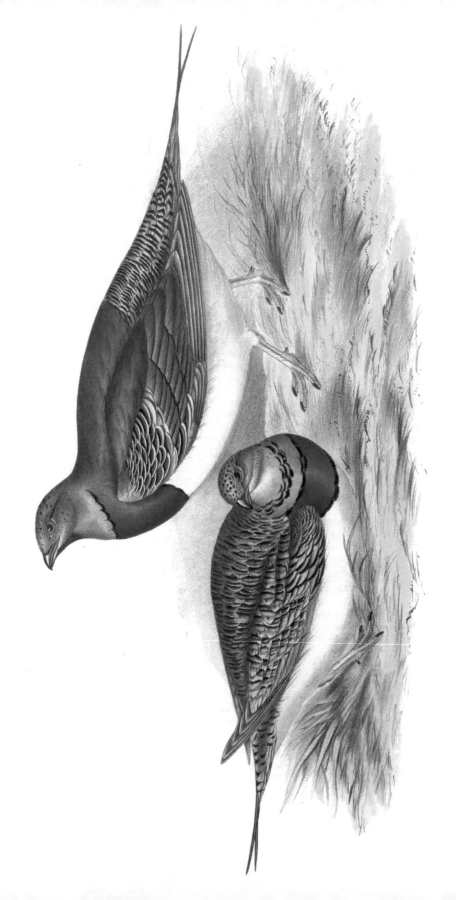

PIN-TAILED SAND GROUSE.

Pterocles Alchata, (Temm.)

PINTAILED SAND-GROUSE.

Pterocles setarius, *Temm.*

Le Ganga Cata.

THE Pintailed Sand-Grouse is a native of the southern portion of Europe, the North of Africa, and the level and arid plains of Persia ; it is also particularly abundant in Spain, Sicily, and through the whole of the Levant, visiting at uncertain seasons, and in small numbers, the southern provinces of France. It is a bird of migratory habits, and, like its congeners, prefers wild and barren districts where the poverty of the soil affords but little inducement to the enterprise of man ; we are consequently unable to obtain any minute details respecting its habits and manners. Its food consists of seeds, insects, and the tender shoots of vege-tables. Its nest, says M. Temminck, is constructed on the earth among loose stones and tufts of herbage, the female being said to lay four or five eggs, the colour of which is unknown. Nothing can be more beautiful, or evince more evident marks of design, than the peculiarities which the great Author of Nature has bestowed upon the birds that compose the great family *Tetraonidæ*, or Grouse, as regards form and colouring in con-nexion with their habits and mode of life. They are all more or less migratory ; but in those species which nature has placed in countries where a luxuriant vegetation supplies them with abundance of food, we find a rounded form of wing, and moderate power of flight, sufficient only to enable them to pass from one pasture or heath to another. It appears also bountifully provided by Providence, that various birds inhabiting countries where the seasons and surface of the earth in summer and winter present striking contrasts, should also undergo a corresponding and analogous change of plumage ;—thus, the different species of Ptarmigan of the northern parts of Europe change their brown livery of summer, which accords so well with the colour of the heathy hills they inhabit, to a pure white in winter, almost rivaling the spotless snow by which they are then for a time surrounded. Their plumage also at this inclement season becomes thicker, and invests the whole of the body even to the extremity of the toes.—If from this we turn to the bird before us, we find an equal provision for its wants and mode of life, varied according to the almost opposite circumstances in which it is placed. Not inhabiting moors or districts covered with verdure, but dwelling in extensive sandy plains, with here and there only a patch of scanty vegetation, and where the season and soil preserve an almost complete uniformity of temperature and appearance, greater powers of flight are required and bestowed ; the wings are elongated and pointed, to enable it to pass with facility over immense tracts in its search after food or water, or to change its situation from one district to another ; the colour of the plumage also remains unchanged throughout the year, that it may ever assimilate with the sandy and stony soil where nature has fixed its abode ; the nostrils remain unconcealed, and the tarsi (although exhibiting rudiments of down,) are naked in comparison with the fur-clad feet of its northern relatives. The connexion which such changes and such modifications of structure evince, in reference to the preservation and protection of the species, cannot fail to suggest themselves to the understanding, and need not be insisted on. The colours of the male and female of the Sand-Grouse differ considerably. In the male, the throat is black ; the cheeks light rufous ; across the breast extends a band nearly two inches broad, of a rufous colour, edged above and below with a narrow black line ; the head, neck, back and scapulars olive-green ; rump and tail-coverts barred with black and yellowish ; the small and middle wing-coverts obliquely marked with chestnut and edged with white ; greater coverts olive inclining to ash-colour, each feather being terminated by a black crescent ; the whole of the under surface of a pure white ; the tail-feathers tipped with white ; the outer one on each side edged with white also ; the two middle feathers are long, and pass gradually into slender filaments exceeding the rest by three inches : length between ten and eleven inches, exclusive of the elongated tail-feathers.

In the female, the throat is white ; below this a partial collar of black which reaches only to the sides of the neck, with the broad orange band and black lines common to the male ; the whole of the upper part barred with black, yellow, and ash-blue ; all the wing-coverts bluish ash ; the primaries have a band of red and terminate with black bars ; the two elongated tail-feathers only exceed the others two inches.

Young birds differ from both parents, in having the general plumage less varied.

We have figured a male and female of the natural size.

EUROPEAN FRANCOLIN.
Francolinus Vulgaris.(Brisé.)

Genus FRANCOLINUS.

GEN. CHAR. *Beak* strong, middle size, convex above, and incurved towards the tip. *Nostrils* basal, lateral, half closed by a naked arched membrane. *Tail* moderate, very slightly rounded, and consisting of twelve feathers. *Feet* four-toed, naked; the *tarsi* of the male with strong blunt spurs. *Wings* short.

EUROPEAN FRANCOLIN.

Francolinus vulgaris.

Le Francolin à collier roux.

It is delightful to examine the series of affinities by which the natural groups of animated nature are connected; and although we cannot at all times clearly trace this connexion, yet we must believe the deficiency to exist in ourselves, and not in those laws of relation which the Creator appears to have impressed upon all his works. In the bird before us we trace, or fancy we can trace, one of those unions through which the splendid-coloured Pheasants of the East are united to the sober-coloured Quails and Partridges of the European Continent; its form and habits connecting it with the latter, while its colouring manifests a relationship to the beautiful Oriental genus *Tragopan*, of which many examples have recently come under our immediate inspection. The near relationship which we fancy exists between the genera *Francolinus* and *Tragopan*, consists in their general style of colouring, in their short spurs, and in the conformation of the beak. Another section of the genus *Francolinus*, peculiar to Africa, exhibits also a form differing from these in the structure of the beak, in which particular, as well as in the uses to which it is applied in obtaining food, it assimilates to the Oriental genus *Lophophorus*: still between these groups we may yet expect to find others, harmonizing with each, so as to form a complete concatenation.

Of the genus *Francolinus*, the present is the only species indigenous to Europe. Unlike its African congeners, which feed on bulbous roots, for procuring which their beak is expressly adapted, our European bird differs little in its food and form of beak from the true Partridge. It however exhibits a preference for moist and humid districts, and perches on trees. In Europe, its habitat appears to be exclusively confined to the southern regions, as Sicily, Malta, and the Neapolitan territories; but it is also found in the North of Africa, and over the greater portion of the Asiatic Continent, and we have also received it in collections from the Himalaya Mountains.

Respecting its habits and nidification we have nothing to communicate. Like most gallinaceous birds, its flesh is very delicate, and much esteemed in India.

In the male, the feathers on the top of the head are black with a margin of yellowish brown; ear-coverts white; circle round the eyes, lower part of the cheeks, sides of the head, and throat of a deep black; below which a broad chestnut collar extends round the neck; wings and back yellowish brown, each feather having a dark reddish brown centre, except those of the quills which are barred with this colour; rump and tail-coverts white, barred with black, as are also the middle tail-feathers, the outer ones being entirely black; breast and lower parts black; sides blotched with black and white; thighs brownish barred with black; under tail-coverts chestnut; beak black; legs reddish flesh-colour; tarsi spurred.

In the female, the general ground colour of the plumage is a yellowish brown, darker on the cheeks and quill-feathers, and becoming paler on the under parts; the feathers of the back and wings are marked as in the male; the breast and under surface irregularly crossed with barb-shaped marks of dark brown; the rump and tail-coverts barred alternately with broad marks of obscure brown and narrow lines of white; under tail-coverts chestnut; beak brownish; legs reddish; tarsi unarmed.

We have figured a male and female of the natural size.

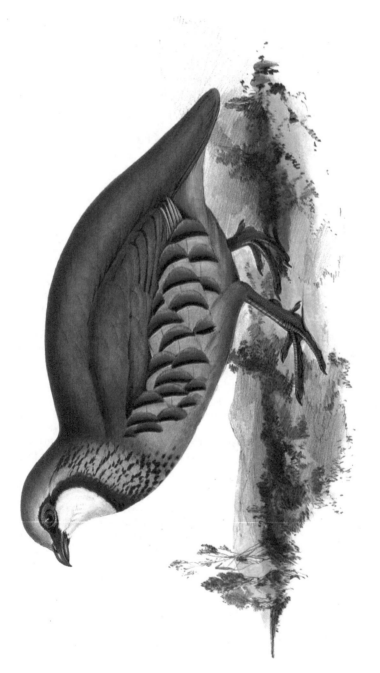

RED-LEGGED PARTRIDGE.
Perdix rubra. (Brifs)

Drawn from Nature & on stone by J. & E. Gould.

Printed by C. Hullmandel.

Genus PERDIX, *Lath.*

GEN. CHAR. *Bill* short, strong, naked at the base; upper mandible convex, with the point bending considerably downwards. *Nostrils* basal, lateral, pierced in a large membrane, and partly concealed by an arched naked scale. *Wings* short, concave, the first three quills shorter than the fourth and fifth, which are the longest. *Tail* of fourteen or eighteen feathers, generally bending towards the ground. *Feet* with three toes before, which are united by a membrane as far as the first articulation, and one behind. *Tarsi* in the male bird frequently with one or more than one spur or tubercle.

RED-LEGGED PARTRIDGE.

Perdix rubra, *Ray.*

Le Perdrix rouge.

FROM this peculiar and beautiful group the Common Partridge may with great propriety be separated, as their habits and characters vary considerably, the Red-legs being spurred, and according to some authors perching on trees, which the common species never does; and it is also destitute of spurs.

Of the Red-legs five distinct species are now recognised, three of which are natives of Europe: of these the species here represented is the most common, and is, we believe, confined entirely to the Continent and the islands of Guernsey and Jersey. Like the Pheasant, the Red-legged Partridge is now naturalized in the British Islands, but it must nevertheless be considered an introduced species, and consequently as not strictly belonging to our Fauna; and we much question whether those who have them on their estates have not cause to regret their introduction, for although highly ornamental in their appearance, their flesh is not equal to that of the Common Partridge (*Perdix cinerea*), which from its diminutive size and less pugnacious habits is compelled to retreat and give place to its more powerful opponent. It is more shy and wary than the common species, and is very difficult to approach, even at the commencement of the shooting-season, a covey being seldom flushed without having run before the dogs for a considerable distance, when they mostly rise out of gunshot. It is now becoming extremely numerous in many parts of England, particularly Suffolk and the adjoining counties. Although it does extremely well in preserved manors, arable lands, &c., still it appears to evince a partiality for sterile wastes and heathy grounds. It is very abundant throughout the plains of France and Italy, is rarely found in Switzerland, and scarcely if ever in Germany or Holland. It is very prolific, the female laying from fifteen to eighteen eggs, of an orange yellow freckled all over with markings of a red colour. The young before the second moult have their plumage striated somewhat after the manner of the young of the common species, but by the end of October this colouring is exchanged for the transversely marked plumage of the adult: the old birds of both sexes are so nearly alike in the colouring and markings, that were it not for the blunt spur, which always forms an appendage to the male, it would be difficult to distinguish them. Its food consists of wheat and other grains, vegetables, insects, &c. Its flesh is whiter and more dry than that of the common species.

The male has the forehead grey; crown of the head and whole of the upper surface greyish brown with a tinge of rufous; throat white surrounded by a black band, which dilates upon the chest and the sides of the neck into a number of small black spots on a grey ground; abdomen and under tail-coverts sandy red; feathers of the flanks grey at their base, to which succeeds a broad transverse band of black, the tips being chestnut red; outer tail-feathers rufous, the centre ones more grey; legs, bill, and eyelids red.

The Plate represents a male of the natural size.

BARBARY PARTRIDGE.
Perdix petrosa.

GREEK PARTRIDGE.
Perdix saxatilis.

BARBARY PARTRIDGE.

Perdix petrosa, *Lath.*

La Perdrix rouge de Barbarie.

ALTHOUGH this species is generally known by the appellation of the *Barbary Partridge*, from its common occurrence on that line of the coast of Africa, nevertheless it is equally frequent in the southern portion of Europe which borders the Mediterranean, and in the islands of that sea; breeding abundantly among the rocky mountains of Spain, and in the islands of Majorca and Minorca, in Sardinia, Corsica, Malta, and Sicily. Its occurrence in France is very rare, and then only accidental, nor is it known to visit the more northern parts of Europe.

In the general character of its plumage, the *Perdix petrosa* bears a striking resemblance to the two other species of Red-legged Partridge, which are also indigenous to Europe, but may at once be distinguished by the rufous brown collar round the neck, thickly spotted with white points. In habits and manners it is strictly identical with the well-known Guernsey Partridge, in the description of which we have entered more fully into the details of the subject. The female chooses barren places and desert mountains, where among low bushes she deposits her eggs to the number of fifteen, the colour of which is yellowish, thickly dotted with greenish olive spots. Grain, and insects occasionally, form, as is the case with the others of the genus. the food of this species.

The beak and a bare space round the eyes are red; the legs, which in the male are furnished with a short blunt spur, are also red; irides hazel; a deep chestnut stripe commences at the gape and runs over the top of the head to the back of the neck, where it passes off on each side, forming a collar round the neck studded with white spots; a broad line above the eyes; the cheeks and throat are of a dull blueish-ash colour, but the ear-feathers are reddish brown. The whole of the upper surface, with the exception of a few blue feathers edged with red near the shoulders, is of a brownish grey. Breast dull ash-colour; the sides barred transversely with ferruginous brown and black on a light ground; each feather barred with grey, black, brown and white, ending in a band of darker ferruginous brown; the under parts light reddish-brown. Tail chestnut. Length thirteen inches.

The female differs only in being rather smaller, the collar round the neck somewhat narrower, the general plumage scarcely so bright, and the absence of spurs on the tarsi.

GREEK PARTRIDGE.

Perdix saxatilis, *Meyer.*

La Perdrix Bartavelle.

OF the three species of Red-legged Partridges which inhabit Europe, the present is the most rare. In size and general colouring it is not unlike its allied congeners, which, with one from the Himalaya Mountains, forms a beautiful group, embodying differences, we think, sufficiently marked to warrant its separation into a new genus, distinct from that of which the Common Partridge of our corn-fields is a familiar example.

The localities of the present species are much more northern than those of the *Perdix petrosa*. It inhabits the Alps, Tyrol, and Switzerland; as well as Italy, the Archipelago, and Turkey; frequenting the higher regions of the mountains during the summer, and descending towards the valleys as winter approaches. M. Temminck informs us that it breeds among the moss and herbage which covers the surface of rocks and large stones, laying fifteen eggs or more, very much resembling those of the preceding species. The beak, the circle round the eyes, and the legs are red; the tarsi armed with a short blunt spur. Irides hazel; a black band beginning at the beak passes through the eye down each side of the neck and meets on the chest, inclosing the cheeks and throat, which are white; the top of the head, the back of the neck, and the whole of the upper parts of the body are of a blueish ash colour, the feathers across the shoulder having a vinous tinge; the breast cinereous; the sides barred as in the preceding species,—with this difference, that the black bands are not so far apart, and the intervening space is of a delicate fawn colour; the lower part of the belly is of a yellowish cream colour; the tail consists of eighteen feathers, of a deep chestnut.

There is no difference between the sexes, with the exception of the female being smaller in size and destitute of spurs.

Our Plate represents a male of each of these species, of the natural size, and in the adult plumage.

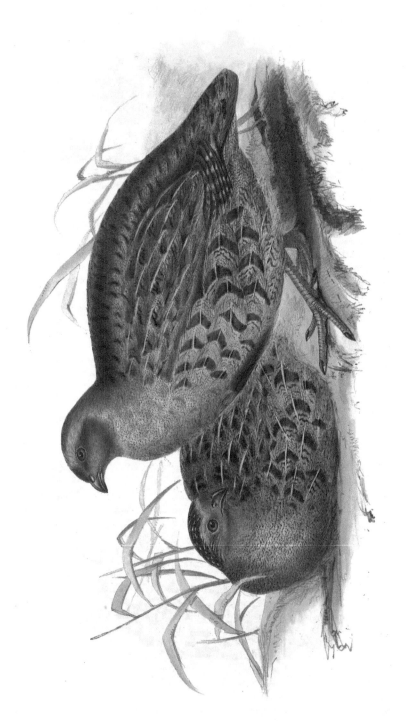

COMMON PARTRIDGE.
Perdix cinerea. *(Lath.)*

Drawn from Nature & on Stone by J & E Gould.

Printed by C. Hullmandel.

COMMON PARTRIDGE.

Perdix cinerea, *Lath.*

La Perdrix grise.

So exclusively European is this celebrated bird, that as far as our own observations go, and these have not been circumscribed, we have never seen an example either from Asia or Africa; although M. Temminck states that it visits Egypt and the shores of Barbary. In affinity it appears to us to rank directly intermediate between the Quails and Redlegs, and with some species from India to form an independent genus to the exclusion of the Quails on the one hand and the Redlegs on the other; and as the Quails have been already separated, it is to us very evident that the Redlegs ought to be separated also.

We do not propose to enter into any details respecting the circumstances that render the Partridge so interesting to sportsmen, as the subject has already engaged the attention of numerous writers, to whose accounts we have nothing to add.

The Partridge pairs early in spring, when fierce contests ensue between the males for the possession of the females. They rear one brood in the year, consisting of from ten to eighteen young, which generally make their appearance about the end of June, and continue associated during the autumn and winter, forming what the sportsman calls a covey, and in the ensuing spring separate, each selecting its mate. The eggs are deposited on the ground in a small hollow, scratched for the purpose under the cover of a tuft of grass or any similar material, and is not unfrequently found in fields of clover or standing corn. The males are distinguished from the females by being larger in size, by possessing a brighter colour about the face, by having a large chestnut-coloured mark on the breast, and by wanting the transverse bars of brown on the upper surface so conspicuous in the plumage of the female.

The Partridge prefers wide tracts of rich corn land to more barren and uncultivated districts, and in bleak and mountainous situations is almost unknown.

The male has the cheek, throat, and a stripe over each eye pale buff; the neck and breast bluish grey ornamented with fine zigzag black lines; on the breast a large horseshoe-shaped patch of chestnut brown; flanks grey, banded with pale brown; back, wings, rump, and upper tail-coverts brown transversely barred and spotted with black; shafts of the scapularies and wing-coverts yellowish white edged with black; quills blackish grey barred with brown; tail reddish orange; bill, legs, and toes bluish grey; irides brown; naked skin behind the eye red.

The particulars in which the female differs having been pointed out above, it will be unnecessary to repeat them here.

We have figured male and female of the natural size.

QUAIL.

Coturnix dactylisonans; (Meyer)

Drawn from Nature & on Stone by J & E Gould.

Printed by C. Hullmandel.

Genus COTURNIX.

GEN. CHAR. *Beak* short and somewhat feeble, the upper mandible curved towards the point. *Nostrils* basal, lateral, and half covered with a membrane. Orbits closely surrounded by feathers. *Wings* moderate, having the first and second quill-feathers the longest. *Tarsi* smooth, without spurs or tubercles. *Toes* four in number, three before and one behind. *Tail* short, rounded, and concealed by the tail-coverts.

QUAIL.

Coturnix dactylisonans, *Meyer.*

La Caille.

WITHOUT commenting upon the propriety of separating the Quails from the Partridges, a point on which we are fully decided, we shall at once enter upon a history of the bird before us. No individual of the Gallinaceous order enjoys so wide a range in the Old World as the Common Quail: it is abundant in North Africa, most parts of India, and, if we mistake not, China; while the whole of the southern portions of Siberia, and every country in Europe except those approximating to the polar circle, are visited by it annually, or adopted for a permanent abode. A considerable number are stationary in the southern portions of Europe, such as Italy, Spain, and Portugal, but their numbers are greatly increased in the spring by an accession of visitors, which emigrate from the parched plains of Africa, in search of more abundant supplies of food, and a congenial breeding-place. So vast and countless are the flocks which often pass over to the islands and European shores of the Mediterranean, that a mode of wholesale slaughter is usually put in practice against them, a circumstance which no doubt tends to limit their inordinate increase. They are polygamous in their habits; and in their migrations the males always precede the females, and are easily decoyed into nets by an artificial imitation of the voice of the latter. This mode of taking them is practised to a great extent in France and other parts of the Continent, which accounts for the vast majority of male birds yearly imported from thence into the London markets. In the British Islands the Quail is more sparingly dispersed, arriving in spring as soon as the tender corn is of a sufficient height to afford it shelter, and remaining with us till it has performed the duties of incubation, when it retires by gradual journeys towards the south; for although when flushed in our fields its flight is neither protracted nor elevated, it is enabled to perform its migrations with greater ease than the general contour of its body would lead us to expect.

The eggs are from eight to twelve in number, of a pale yellow brown blotched and dotted with darker brown and black, and are deposited on the ground with little or no nest.

The sexes may be distinguished by the male having a black mark on the throat, which part in the female is white. The young of the year so closely resemble the female that they are scarcely to be distinguished.

The general plumage of the upper surface is brown, beautifully variegated with dashes of black and yellow, and numerous fine zigzag transverse lines of black; the scapularies and the feathers on the flanks have each a lanceolate stripe of yellowish buff down their centres; the chin is dusky white bordered in the male with black; the breast and belly pale buff, the sides being streaked and mottled with reddish brown, black, and white; tarsi brownish flesh colour; bill brown.

The Plate represents a male and female of the natural size.

ANDALUSIAN TURNIX.

Hemipodius tachydromus. *(Temm.)*

Genus HEMIPODIUS, *Temm.*

GEN. CHAR. *Beak* moderate, slender, straight, very compressed; culmen elevated and curved towards the point. *Nostrils* lateral, linear, longitudinally cleft, partly closed by a naked membrane. *Tarsi* rather long. *Toes* three before, entirely divided: no posterior toe. *Tail* composed of weak yielding feathers clustered together, and concealed by the feathers of the back. *Wings* moderate, the first quill-feather the longest.

ANDALUSIAN TURNIX.

Hemipodius tachydromus, *Temm.*

La Turnix tachydrome.

THE birds of this genus are mostly inhabitants of the intertropical regions of the Old World; two species, it is true, have been discovered in the southern parts of Europe, but in such limited numbers as to prove satisfactorily that the northern portions of Africa are their true habitat, consequently the southern parts of Spain, Italy, and the Islands of the Mediterranean are among the utmost limits of its range northward. They differ from the true Quails (*Coturnix*), in the total absence of the hind toe, and in the long and slender form of their bills: they are the most diminutive birds of the gallinaceous tribe, being not more than half the size of the Common Quail. M. Temminck states that they are polygamous, and that they give a preference to sterile lands, sandy plains, and the confines of deserts, over which they run with surprising quickness; he also states that the young and old do not associate in company or in bevies as is the case with the Quail. Their food is said to consist principally of insects, to which are added small seeds, &c.

The sexes are so much alike that it is very difficult to distinguish them by their plumage. Col. Sykes states that the birds of this genus which he observed in the Dukhun, viz. *Hem. pugnax*, Temm., *Hem. Taigoor*, Sykes, and *Hem. Dussumier*, Temm., were either solitary or in pairs, and mostly found in pulse and *Chillee* fields (*Capsicum annuum*)." The last-named species "frequents thick grass, and sits so close as to expose itself to the danger of being trodden upon," and its "flight is so abrupt and short, that ere the gun is well up to the shoulder the bird is down again," in all which respects we doubt not the bird here represented very closely resembles them.

The Andalusian Turnix is tolerably abundant at Gibraltar and that part of Spain which borders the Mediterranean, being more scarce in the central portions, and in the northern and all similar latitudes altogether absent.

The top of the head is dark brown, streaked longitudinally with reddish yellow; throat white; the feathers on the sides of the chest reddish chestnut, those of the flanks yellowish white, with a crescent-shaped mark of rich brown occupying the centre of each; lower part of the belly pure white; the upper surface is dark brown with numerous zigzag lines of reddish ash, and transversely rayed with lines of brown and chestnut, each feather being finely margined with white; coverts of the wing yellow with a spot of reddish chestnut on the inner web; primaries ashy brown, the outer web bordered with white; bill and legs greyish flesh colour.

The Plate represents a male and female of the natural size.

COLLARED PRATINCOLE.
Glareola torquata. (*Meyer*)

Genus GLAREOLA.

GEN. CHAR. *Beak* short, hard, convex, curved for upwards of half its length, and compressed towards the point. *Nostrils* at the sides of the base, oblong, and obliquely cleft. *Legs* feathered nearly to the knee; toes three before and one behind, the outer united to the middle one by a short membrane; claws long and drawn to a fine point. *Wings* very large, the first quill-feather the longest. *Tail* more or less forked.

COLLARED PRATINCOLE.

Glareola torquata, *Briss.*

Le Glaréole à collier.

THE genus *Glareola* appears to be strictly confined to the Old World, no Transatlantic example having ever been discovered; nor, indeed, are we aware of any form in the ornithology of America which at all approaches the present. Three species are all that are as yet discovered. Of these, two (the *G. grallaria* and the *G. lactea*) are peculiar to the eastern provinces of Asia and Africa; the other, the bird now before us, is spread throughout the warm and temperate regions, not only of these continents, but Europe also: hence it would seem as if Nature endeavoured to make up by extent of habitat for the limitation of species. Still, however, although thus diffused, the Pratincole may be said to be truly a native of the eastern provinces of Europe, on the Asiatic borders, and especially Hungary, where wide tracts of morass and flatlands, abounding in lakes both fresh and saline, and traversed by mighty rivers, afford it food and security. "In Hungary," says M. Temminck, "among the immense morasses of the lakes Neusidel and Balaton, I have been in the midst of many hundreds of these birds;" and we may add, that it is no less abundant in Western Tartary. In England it is only an occasional visiter; but in Germany, France and Italy it is a bird of periodical occurrence.

With the long wings and forked tail of the Swallow, the Pratincole possesses that rapidity and power of flight for which that bird is so remarkable. It takes its food, which consists of insects, and especially such as frequent marshes and the borders of rivers, while on the wing, darting along in the chase with the rapidity of an arrow; nor is it less distinguishable for celerity on the ground, and often catches its prey as it nimbly runs along.

This elegant and graceful bird incubates in the concealment afforded by reeds, osiers and tall herbage, laying three or four white eggs.

As it respects their plumage, the sexes offer no difference. The young are more obscure in their tints, the upper parts being clouded with dull brown, and the throat being dirty white.

The adult plumage may be thus detailed. Head, back of the neck, and whole of the upper surface greyish brown, except the secondaries, which are tipped with white, and the upper tail-coverts and the lower portion of the tail-feathers, which are white; quills and remaining portion of the tail-feathers blackish brown; throat and sides of the cheeks white with a wash of buff, and bordered by a narrow black band, which takes its origin beneath the eye; under wing-coverts deep ferruginous; chest brownish grey passing into pale fawn or white; abdomen dull white; beak black, except the base, which with the irides and circle round the eye are reddish brown; tarsi brownish ash colour.

Our Plate represents a male and female of their natural size.

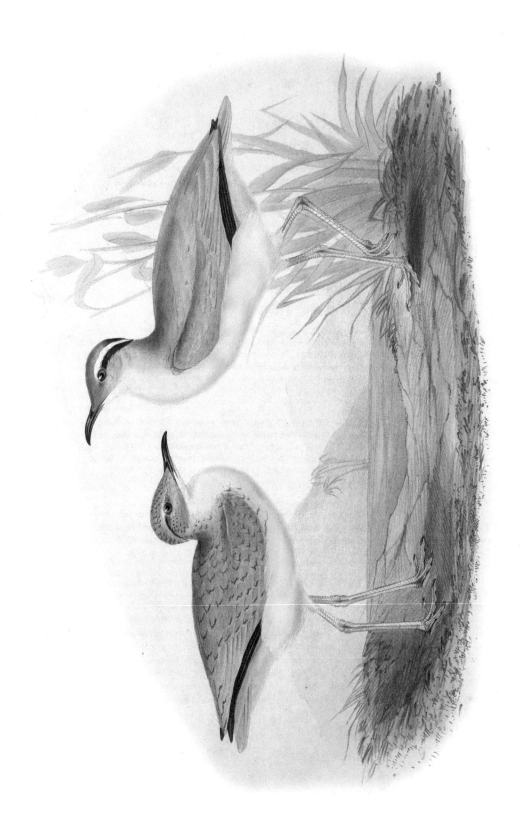

CREAM COLOURED COURSER.
Cursorius Isabellinus. *(Meyer)*.

Genus CURSORIUS, *Lath.*

GEN. CHAR. *Beak* shorter than the head, depressed at its base, slightly convex at its points, and somewhat curved and pointed. *Nostrils* oval, surmounted by a small protuberance. *Tarsi* long and slender. *Toes* three before only, entirely divided, the inner toe scarcely equalling half the length of the middle toe. *Wings* long; the first and second quill-feathers nearly the same length, and the greater coverts as long as the quill-feathers.

CREAM-COLOURED COURSER.

Cursorius Isabellinus, *Meyer.*

Le Court vite Isabelle.

THE most superficial view of the structure and proportions of this singular and elegant bird, would at once enable us to perceive that it is a fleet and rapid courser, peculiarly fitted for dry and sterile situations, such as sandy deserts, plains, and open tracts of country.

Although in many respects it approaches the Bustards, yet it has with great propriety been separated from that group into a distinct and well-defined genus, which now includes five or six species, closely united to each other in form and general habits, and, as well as the genus *Otis*, strictly confined to the older known portions of the globe. Africa supplies some peculiar species, continental India one or two others ; and we have seen one from the islands of the Indian Archipelago differing from every other.

The present species, the *Cursorius Isabellinus*, although frequently visiting Southern Europe, and occasionally our own island, is, strictly speaking, a native of Northern Africa and Abyssinia, where, from the remarkable similarity of the colour of its plumage, it finds security among the sandy deserts and plains, from which it can hardly be distinguished, and where it speeds along with the swiftness of an arrow, in pursuit of its food, which consists principally of the insects peculiar to such situations. Of its nidification no authentic information has reached us ; but most probably, like the Bustard, it incubates among loose stones and the scanty herbage of the desert.

We have had the good fortune to obtain examples of the young, in addition to the adults, of both sexes, which latter do not offer any material difference of plumage. The young, although nearly arrived at its full size, may be distinguished by the absence of the black occipital patch, as well as the bands of black and white which pass from behind the eye, and by its plumage being obscurely waved with dusky, transverse and somewhat arrow-shaped markings.

The general plumage of the adult is of a delicate fawn colour above, lighter beneath, the occipital and quill-feathers being black ; from above the eye to the back of the neck extends a white line, circumscribing the black portion of the occiput, beneath which, from the angle of the eye, runs a similar, but still narrower band of black ; tail, with the exception of the two middle feathers, banded near its extremity with black, and tipped with dull white ; beak black ; legs light cream colour.

The Plate represents an adult male, and a young bird of the first autumn, both of the natural size.

GREAT BUSTARD.
Otis tarda (Linn.)

Genus OTIS.

GEN. CHAR. *Bill* moderate, straight, depressed at the base, and having the point of the upper mandible curved. *Nostrils* removed from the base, lateral, oval, and open. *Legs* long, naked above the knee. *Tarsi* reticulated. *Toes* three, all forward, short, united at the base, and bordered with membranes. *Wings* of mean length, the third quill-feather the longest.

GREAT BUSTARD.

Otis tarda, *Linn.*

L'Outarde barbue.

As man extends his dominion over the globe, many races of animals, and among them the present family, will gradually disappear from the haunts where they now abound, and linger only in localities which hold out no inducements to the exertions of human enterprise. Africa then, we may venture to predict, will afford on its extensive plains the last asylum in which the Bustard may find security, till at length, like the Dodo, many species of this stately family will become extinct, leaving only their remains for our investigation.

So nearly has the Great Bustard become annihilated in the British Islands that it is even doubtful whether any males still exist to accompany the few old females which remain on some of the extensive inclosures and large fields of turnips in the counties of Norfolk and Suffolk. On the Continent, as might be supposed, with the exception of Holland, the Great Bustard is tolerably numerous; a circumstance to be accounted for by the comparative thinness of the population and by the wide and extended plains which there exist. From the sandy deserts of Spain and Italy, which afford a congenial nursery wherein it may dwell and rear its young in safety, its range extends as far north as Siberia and Kamtschatka, but no example of this genus has ever been seen in America.

The eggs are two in number, of an olive green colour with slight darker variations; they are rather larger than those of the Turkey, and are deposited on the bare ground among the herbage, frequently in clover, trefoil, and corn. The young are hatched in a month, and from the day on which they are excluded from the shell they follow their parents over the plain until the following spring. The Great Bustard rarely takes wing unless so hard pressed as to have no other chance of escape, but runs with great strength and swiftness for the distance of several miles.

As an article of food the flesh of the Bustard is highly esteemed, and on the Continent the bird is frequently to be seen exposed in the markets for sale.

The food of this bird consists of various kinds of grain, to which are added the tender leaves and shoots of trefoil and other vegetables; nor does it disregard snails, insects, mice, &c.

The male has a large and membranous stomach, and possesses a lengthened gular pouch, extending down the fore part of the neck, the entrance of which is situated beneath the tongue; although the purpose of this pouch is not yet clearly ascertained, it is supposed by some to be formed for holding water, with which it might supply itself, the female, or young during the period of incubation; "yet," says Mr. Selby, "this supposition does not carry with it much probability, as the male is never seen in close company with the female except previous to the time of laying;" thus leaving us still in doubt what precise purpose this singular organ is designed to perform.

The general weight of the male when in good condition is from twenty-eight to thirty pounds, and when fully adult, which is not until it is at least five or six years old, may be distinguished from the female not only by its larger size, but also by the lengthened wiry feathers which spring from the sides of the face and extend backwards for several inches, and by the rich band of dark grey which crosses on the chest from side to side.

The head, nape, and fore part of the neck and chest fine ashy grey; a streak of brown passes from the top of the head to the occiput; chin, throat, and mustaches white; lower part of the neck and sides of the chest rich chestnut brown inclining to reddish orange, strongly barred, variegated with black and grey; secondaries and greater coverts grey; quills black; tail-feathers white at their base and tip, the intermediate space being barred with black and reddish brown; belly and vent white; legs brownish black; beak bluish grey.

The female resembles the male in the colour of her plumage, but differs from him in wanting altogether the mustaches and in being only about half his size.

The Plate represents an adult male and female rather less than half the natural size.

RUFFED BUSTARD.
Otis bruchara. *Linn.*

RUFFED BUSTARD.

Otis Houbara, *Linn.*

L'Outarde Houbara.

ALTHOUGH the present beautiful species of Bustard has been occasionally killed in Spain and other parts of Europe, its native habitat is undoubtedly Arabia and Northern Africa, where extensive sandy deserts afford it a situation congenial to its natural habits. Dr. Latham, in his "General History of Birds," informs us, that according to Bechstein the Houbara has been killed in Siberia; but we have now some reason to doubt this conclusion, as J. E. Gray, Esq., of the British Museum, has lately made known a species of Bustard from the elevated range of the Himalaya which extends itself into Siberia, and we therefore suspect Bechstein has confounded this bird with the *Otis Houbara*, as it is characterized by the same singular kind of ruff and general style of colouring; although it may be distinguished from it by its inferiority in size and other minor particulars. To his bird Mr. Gray has given the specific name of *Macqueenii*; and the circumstance of its discovery is the more interesting, as we have now two species of Bustards exhibiting this singular disposition of plumage, which at once distinguishes them from the other birds of that family.

Athough so rarely met with in Europe, the Houbara abounds in Africa, where it is much prized for the excellence of its flesh, which is considered one of the greatest delicacies, and is said to be of exquisite flavour.

The history of this bird is at present but imperfectly known, European naturalists being unacquainted with its eggs or nidification : nor have the characters of the female been yet observed; so that we are unable to say whether or not she possesses that ornamental plumage which graces the male; most probably if the feathers of the ruff exist at all, they are much less perfectly developed; and indeed, as it regards the male, we have yet to learn whether he does more than possess these long feathers during the breeding season, and lose them subsequently, a change analogous to that which we know so frequently occurs in others of the feathered race.

The beak is lengthened and depressed at its base, and, together with the feet, is of an olive colour. The head is surmounted by a crest of long and slender filamentous feathers of a pure white; the top of the head, the cheeks, occiput, back, and forepart of the neck grey, with minute zigzag bars and spots of brown; from the sides of the neck spring two large tufts of flowing feathers, gradually increasing in length to the extent of seven or eight inches; the upper portion of which is black, the remainder white; the whole of the upper surface is light tawny, each feather being irregularly marked with transverse zigzag bars of brown; the primaries dark brown at the tip, and white at their base; the tail-feathers besides being spotted are ornamented with three bands of blueish grey; the breast and under parts pure white. Total length from twenty-five to twenty-eight inches.

Our figure is that of an adult bird, two thirds of its natural size.

LITTLE BUSTARD.
Ous tetrax. *(Linn.)*

LITTLE BUSTARD.

Otis tetrax, *Linn.*

L'Outarde canepetiere.

ALTHOUGH the Great Bustard (*Otis tarda*, Linn.) was at one time common in England, we are by no means so well assured that such was also the case with respect to the bird before us; indeed we should suspect, from the localities which it affects, that its visits to the British shores have ever been, as at the present day, accidental and of rare occurrence. Its habitat appears to be more exclusively confined to the southern portion of Europe, especially Spain, Italy and Turkey, as well as the northern coast of Africa; and although occurring in the central parts of France, it is by no means a common bird; nor is it at all found in the northern parts of the European Continent. If, however, we are not to consider the Little Bustard as one of the birds strictly indigenous to our island, still the circumstance of its having been often killed in England fully entitles it to a place in the Fauna of this country. Of the various British specimens taken, we may refer among others to one in the possession of His Grace the Duke of Northumberland, shot at Warkworth, in the autumn of 1821; another in the possession of Mr. Selby, shot in February 1823; and a third in the collection of Mr. Yarrell, which was taken near Harwich.

The specimens above enumerated, as well as all those of which any report has reached us, have been either invariably females or immature males, and in no instance an adult male, so conspicuous for the beautiful and singular markings which ornament the plumage of his neck and chest. We may here observe, that we have been unable satisfactorily to ascertain, either from our own observation or the information afforded by M. Temminck, whether these bold and decided markings constitute its summer plumage, being lost during winter, as in many species of the allied genus *Charadrius;* or if they are borne throughout the year, so as to constitute a permanent characteristic. In the specimen (in the author's collection) from which our figure was taken, this beautiful state of plumage is exhibited in a manner which the pencil is hardly adequate to convey.

The habits and manners of the *Otis tetrax* are strictly characteristic of the genus to which it belongs; and its general conformation and strength of limb render it well adapted for the station it occupies among those birds whose province is more peculiarly the ground, the surface of which affords them food and a place for nidification. The present species frequents open and extensive wilds or uncultivated districts, particularly uncovered arid plains, where, far removed from the habitation of man, it finds a secluded abode consonant to its reserved and timid disposition; and in these places, among the short herbage, it constructs an inartificial nest, and deposits from three to five eggs, of a uniform glossy olive-green.

In the male, the top of the head and occiput are light yellow, contrasted with numerous dots and lines of black and brown; the throat and cheeks slate-colour deepening, as it proceeds, to black, which continues in a line for some distance down the front of the neck, around which runs a necklace of pure white, commencing on each side of the occiput; the back of the neck (where the feathers are elongated into a short mane), and the sides, are of a deep jet black which meets across the lower part of the neck beneath the white necklace; over the breast extends a large crescent-shaped collar of white, below which is a narrower one of black; the whole of the back and sides of the chest light yellow with shades of reddish brown, thickly barred and dotted with elegant zigzag markings of black (which follow the outline of each feather,) interspersed, especially about the upper part, with large black spots and dashes; the edges of the greater wing and tail-coverts white; the quill-feathers blackish brown; tail yellowish with zigzag markings like those of the back, and crossed by indistinct bars, with indications of others; the middle of the chest and whole of the under surface pure white; bill olive-brown; irides orange; legs and tarsi yellowish grey.

Length eighteen inches; tarsi three inches; middle toe one inch and a quarter. From the joint of the tarsus to the feathery part of the thigh one inch; wing lengthened, and somewhat rounded.

The females and young males have the whole of the upper surface barred as in the male with dark brown zigzag markings on a fawn-coloured ground; the wing-coverts edged with white; the quill-feathers dark brown; the chin white; neck yellow, marked with longitudinal stripes, which as they proceed merge into transverse bars, becoming more and more decided on the chest, where the ground is still yellow; the under parts are pure white.

We have figured a male and female in their full plumage, two thirds of their natural size.

COMMON CRANE.

Grus cinerea. *Bechst.*

E. Lear del et lithog.

Printed by C. Hullmandel

Genus GRUS.

GEN. CHAR. *Beak* longer than the head, straight, strong, compressed, pointed. *Nostrils* placed horizontally in the anterior part of a furrow, large, concave, pervious, posteriorly closed by a membrane. *Legs* long, strong, naked above the joint; three toes in front; middle toe united to the outer one by a membrane; hind toe articulated high up on the tarsus. *Wings* moderate, rounded; first quill-feather shorter than the second; the third the longest.

COMMON CRANE.

Grus cinerea, *Bechst.*

La Grue cendrée.

THAT the Crane was once common in England is a fact learned from the accounts of all the writers on Falconry, who enumerate it among the noblest game, which the Jer and Peregrine Falcons could alone encounter. The gradual cultivation of the country, the draining of marshes, and the inclosure of wild tracts since those days, have almost wholly banished this elegant bird from our island; still, however, it pays occasional visits, and few seasons pass without a specimen being killed within the precincts of England. As in the present day, it must then have been a bird of passage, appearing only in autumn and winter; since its native climate appears to be the higher northern latitudes, both of Europe and the adjoining parts of Asia, whence they pass southwards, being forced to abandon their solitary realms upon the approach of winter, and gladly returning when spring opens the frozen regions, and again offers a friendly asylum.

Flocks of these birds are seen at stated times in France and Germany, passing northwards and southwards as the season may be, in marshalled order, high in the air, their sonorous voices distinctly sounding even from their elevated course. Occasionally they descend, attracted by new-sown fields, or the prospect of finding food in marshes, the borders of rivers, or even the shores of the sea; but generally they continue their flight unchecked towards their destined resting-place.

The food of the Crane is of a more mixed nature than is usual among the great class of Waders, grains and plants, especially such as grow in morasses and moist lands, being added to worms, frogs and fresh-water shells.

The nest is usually placed among reeds, thick osier beds, and the matted foliage which borders lakes and morasses; but sometimes also on the tops of old ruins and similar buildings, where solitude invites to the great task of incubation. The eggs are two in number, of a dull greenish hue with dashes of brown.

The young of the year, besides having the plumes of the wings little developed, are distinguishable by the want of the bare space on the top of the head, or at least in its being but barely indicated, while the black of the front of the neck and occiput is not yet apparent, or indicated only by a few dark streaks.

The adult birds, male and female, are similar in colour, the plumes being less elongated and graceful in the female.

The whole of the body is of a delicate grey, the throat, the fore part of the neck, and the occiput, being of a deep greyish black; the forehead and space between the eye and the beak garnished with black hairs; the top of the head is naked and red; the secondaries form a beautiful flowing pendent plume, each feather being long and decomposed, consisting of loose unconnected barbs hanging half way to the ground; the beak greenish black, passing into horn colour at the tip, but reddish at its base; tarsi black; irides reddish brown. Length, from beak to tail, three feet ten inches.

The aged birds have a white space passing from behind the eye over the cheeks, and along the side of the neck for a considerable distance.

Our Plate represents an adult male nearly one half of the natural size.

WHITE CRANE.
Grus leucogeranus. *(Temm).*

WHITE CRANE.

Grus leucogeranus, *Temm.*

La Grue leucogerane.

This splendid species of Crane having been lately added to the European Fauna, we have deemed it necessary to include it in the present work, and have accordingly figured it from a beautiful specimen of the male presented to us by M. Temminck, who in a letter accompanying it states that it is one of our most recent accessions, and is consequently one of the rarest among the European birds. Its native habitat is doubtless the northern and central portions of Asia, whence its range is extended even to Japan, where it is common. The only European localities in which it has as yet been observed are the most easterly portions of the Continent.

In size this species exceeds the Common Crane, and independently of its snow-white colouring, it also differs from that species in its much longer bill.

Snails, frogs, the fry and ova of fishes, small crustacea and bulbous roots are said to constitute its food.

The whole of the plumage, with the exception of the primaries, which are brown, is of a pure white; bare part of the head red; bill greenish horn-colour; legs and feet black.

The figure is about one third of the natural size.

NUMIDIAN DEMOISELLE.

Anthropoides Virgo. (Vieill.)

Genus ANTHROPOÏDES, *Vieill.*

GEN. CHAR. *Bill* scarcely longer than the head, entire above, sulcated. *Nostrils* linear. *Head* either feathered, or the temples naked. *Feet* four-toed, cleft; the outer toes connected by a membrane at the base.

NUMIDIAN DEMOISELLE.

Anthropoïdes Virgo, *Vieill.*

La Grue Demoiselle.

AFRICA is undoubtedly the true habitat of the members of this genus, of which the bird here figured is a typical example; at the same time, that the range of this species is exceedingly extensive is proved by the circumstance of our having lately seen a specimen killed in Nepâl, and we are of opinion that it is also sparingly dispersed over other parts of India. In Africa the Numidian Demoiselle is abundant over the whole of its northern portions, particularly in the neighbourhood of Bildulgerid and Tripoli, while its beautiful ally the *Anthropoïdes Stanleyanus* of Mr. Vigors is confined to its southern portions. Dr. Latham mentions that it is very common along the whole of the African coast of the Mediterranean; we need not therefore be surprised at its being included in the Fauna of Europe, as its great powers of flight would readily enable it to cross the Mediterranean; and this we find to be the case, as the same author informs us it is found in the southern plains about the Black and Caspian Seas, that it is frequently seen beyond Lake Baikal, about the rivers Selinga and Argun, but that it is never seen further north. M. Temminck also includes it in the Fauna of Europe as an occasional visitant to the southern parts of the Continent. It everywhere evinces a partiality for marshes and the neighbourhood of rivers, and feeds upon snails, aquatic insects, small fish, and lizards.

Like the Cranes in general, it bears confinement extremely well, and forms a highly ornamental and docile creature in the menagerie. It has been known to breed in captivity, but of its nidification in a state of nature nothing is on record.

The sexes are alike in plumage.

Cheeks, throat, front part of the neck, primaries, and the tips of the prolonged scapularies black; a tuft of feathers proceeding backward from the eye pure white; crown of the head and all the remainder of the plumage delicate ash grey; bill black at the base and yellow at the tip; legs brownish black.

We have figured an adult male rather more than half the natural size.

COMMON HERON.

Ardea cinerea, (Lath:)

E. Lear del et lithog   Printed by C. Hullmandel

Genus ARDEA.

Gen. Char. *Bill* long, strong, straight, compressed in a lengthened cone; upper mandible slightly channelled, ridge rounded. *Nostrils* lateral, basal, slit lengthwise in the groove, and half shut by a membrane. *Legs* long, slender, naked above the knee. *Toes* three before, the two outer united by a membrane, the inner divided, and one behind placed interiorly. *Claws* long, compressed, sharp, the middle one denticulated on the inside. *Wings* of middle size, the first quill a little shorter than the second and third, which are the longest.

COMMON HERON.

Ardea cinerea, *Linn.*

Le Héron.

Our large sheets of water adjacent to woods and wild scenery have their pictorial beauties greatly enhanced by this majestic bird, which, however, preserved with the utmost rigor when the art of Falconry was in vogue, is now in less estimation; and notwithstanding its preservation at the present period in many extensive heronries, we have every reason to believe its numbers are yearly diminished; circumstances to which the draining of our fens and marshes doubtless also materially contribute: and it is a question whether at no long distant date the Heron will not become as scarce in our island as the Bustard. It may not be out of place here to observe, that the Crane and Stork were once as common with us as the Heron is now; but being migratory birds, and finding that year after year their subsistence decreased, it is not so much to be wondered at that they should altogether abstain from visiting such an inhospitable abode. The range of the Heron is very extensive, being distributed not only over the continent of Europe, but in nearly every portion of the Old World; and in America we find its place supplied by a species closely resembling it in colour and habits, but nearly a third larger.

The Heron is very nocturnal in its general habits, and will when undisturbed remain during the day perched upon the branch of some large tree, where it sleeps away till evening calls it to exert itself in procuring the necessary supply of food; when there are no trees in the neighbourhood, it may be observed reposing in the centre of the marsh, but generally so elevated as to command a view of every approach, so as to be as safe as circumstances will admit. In the dusk of evening it may be seen leaving its retreat, and winging its way to the accustomed fishing-place, where it spends the whole of the night and morning in watching for its prey: to this end it wades into the water, there remaining motionless as a statue, its keen eye watching the approach of its victims,—fish, particularly eels which are working their way into the shallows in search of their own food: let one come within the range of its neck, which is retracted upon its shoulders in readiness for a blow, and quick as lightning it is seized by the never-failing stroke of its sharp-pointed bill. We may here notice the strong digestive powers with which this bird is provided, whence arises the necessity of an exuberant supply of food; and as its means of procuring it are in conformity with its wants, few birds make such destructive havoc in the preserves of fish: in addition to fish, however, it greedily devours frogs, aquatic insects, water rats, mice, &c. Few birds are more buoyant than the Heron; it elevates itself to a considerable height, and is also capable of maintaining its flight over large rivers and tracts of country.

On the earliest approach of spring, these birds assemble in flocks at the accustomed breeding-places, or heronries as they are termed, and either repair the nests of the preceding year, or construct new ones. They are large, flat structures, composed of sticks and twigs, the interior being lined, according to Mr. Selby, with wool and other materials, and are placed on the topmost branches of trees of the highest growth. The eggs are four or five in number, of a blueish green colour. The young are easily reared, and become quite domesticated in captivity, forming stately ornaments to sheets of water in the vicinity of mansions. During the first year of their existence, they are destitute of the flowing plumes of the back, chest and occiput, the whole plumage having an obscure and dusky tinge.

The adult male has the forehead, the sides of the head, throat, and shoulders of the wings, breast and belly pure white; on the front of the neck a double row of oblong spots of black; several long plumes of white springing from the lower part; a long plume of white feathers rising from the occiput; the sides of the chest black; the whole of the upper surface of a silvery grey; the scapularies elongated, forming loose, streaming, pendent feathers which fall over the wings; the beak and the band round the eyes beautiful yellow, with a tinge of grey; and the tarsi olive green.

The female resembles the male externally, but is somewhat smaller.

We have figured an adult male nearly three fourths of the natural size.

PURPLE HERON.
Ardea purpurea: (Linn)

E. Lear del et lith.   Printed by C. Hullmandel.

Edwd Lear del.

PURPLE HERON.

Ardea purpurea, *Linn.*

Le Heron pourpré.

In this elegant species we cannot fail to remark one of those beautiful gradations of form uniting proximate groups which the ornithologist meets with so continually in his survey of the feathered tribes, and which serves to show that the harmony obtaining throughout all great groups is interminable, except by the accidental annihilation of species. These observations apply with peculiar force to the bird before us, which seems to take an intermediate station between the Common Heron on the one hand and the Bittern on the other; to the former it assimilates in the length and slenderness of the neck, in the occipital plumes, and in the lengthened form of the bill, while by its large spreading toes, straight long nails and shorter legs, it is closely connected with the Bittern, to which it also bears a striking similarity in its habits and manners. Unlike the Common Heron, which prefers open countries and the exposed edges of large sheets of water, the Purple Heron haunts the dense coverts of reed-beds, morasses, and swampy lands, abounding in luxuriant vegetation, among which it is concealed from observation, and instead of building its nest on the topmost branches of the tallest trees, it incubates on the ground amongst that herbage which affords it an habitual asylum. As is also the case with the Bittern, the eggs are three in number, and of an uniform pale bluish green.

The range of this species is so great, that we may say in few words it inhabits the whole of Europe, Asia, and Africa. It is especially abundant in Holland, and in the low marshy districts of France; in the British Islands it must be considered as an accidental rather than a regular visitant, and we suspect that many of those killed in England had escaped from captivity, since numbers are annually brought alive from Holland to the London markets, where we have frequently seen a dozen at one time, together with Spoonbills, Common Herons and Bitterns, all in the most beautiful state of plumage, having been captured during the season of incubation; and often accompanied by hundreds of their eggs. We fear that this wholesale traffic has much diminished the numbers of these species, for the supply has been much less abundant during the last two or three years than it was formerly.

The food of the Purple Heron consists of fish, frogs, mice, and insects.

The sexes are alike in plumage after they have attained complete maturity; and may be thus described:

Crown of the head, occiput, occipital crest, a stripe down the back of the neck, another from the corner of the mouth to the back of the neck, and one passing down each side of the neck, black; throat white; sides and front of the neck rufous, the feathers on the lower half of the latter part lighter and with a broad stripe of blackish brown down the centre; the plumes at the bottom of the neck long, acuminate, and of a greyish white; lower part of the back of the neck, back, wings, flanks, and tail bluish grey, tinged with rufous; shoulders and under wing-coverts rich rufous; breast, all the under surface, and the long filamentous ends of the scapularies, deep reddish brown, intermingled with bluish grey; thighs pale rufous; bare space before the eyes and the bill fine gamboge yellow, with the exception of the ridge or culmen, which is brown; irides pale yellow; legs and feet greenish black.

The young are destitute of the occipital crest, and of the elongated feathers of the scapularies and at the base of the neck, until they are three years old; the forehead and crown of the occiput are grey with a reddish tint; the neck is much paler, and destitute of the black stripes; front of the neck white, with longitudinal black spots; under surface reddish white; upper mandible blackish brown; under mandible, bare space before the eyes, and irides pale yellow.

We have figured an adult male about two thirds of the natural size.

SQUACCO HERON.
Ardea comata; (Pallas).

Drawn from life & on Stone by J. & E. Gould.

Printed by C. Hullmandel.

SQUACCO HERON.

Ardea comata, *Pallas.*

Le Heron crabier.

THE Squacco Heron is one of those birds whose occasional visits have procured for it a place in the Fauna of Great Britain. We are in possession of several facts relative to its capture in different parts of England, but more particularly Norfolk and Lincolnshire, and within the last few years we know of more than one killed in the marshy districts surrounding Great Yarmouth; in addition to which Dr. Latham mentions two or three instances of its capture, one in Wiltshire by Mr. Lambert, and another which became entangled in a fisherman's net, whilst it was spread for drying, at Ormsby in Norfolk. The native locality, however, of this beautiful species appears to be along the western confines of Asia, extending into Turkey, the Islands of the Archipelago, and Italy, where it inhabits the banks of stagnant waters, morasses, the sides of rivers, and the low lands on the sea shore; it also visits some parts of Germany, Switzerland and France, but has never we believe been known to extend its migrations to the more northern regions of Europe.

The total length of the present species is about sixteen inches. The top of the head is ornamented with long yellowish feathers having marginal stripes of brown; these feathers become much more elongated at the occiput, whence spring seven or eight long slender flowing plumes of a yellowish white edged with black; throat white; neck, back, and the long filamentous feathers which rise from it, of a tawny yellow; these feathers, however, sometimes assume a vinous tinge; beak azure-blue at the base, passing through greenish white, to its tip which is black; the naked skin around the eyes and the feet of a greenish olive; irides fine light yellow.

The adult males and females differ but little.

In the young, the upper parts of the body and the scapulars are tinged with brown more or less decided; the beak and legs are not so brilliant in their colouring, but more inclined to a dull yellowish olive; nor do they assume the long occipital plumes or the lengthened feathers which ornament the head and back of old birds in their perfect livery, until they are more than two years old.

According to M. Temminck the nest is built in trees; but nothing is known respecting the colour or number of its eggs.

It subsists on small fish, frogs, marine insects, and mollusca.

We have figured a male in its full plumage, somewhat less than the natural size.

GREAT EGRET.
Ardea alba (Linn.)

E. Lear del et lith.

Printed by Hullmandel.

F. Lear.

GREAT EGRET.

Ardea alba, *Linn.*

Le Heron Aigrette.

This beautiful species of Heron is an inhabitant of the eastern and southern parts of Europe, and the adjacent portions of Africa and Asia. It is said to be found in abundance on the shores of the Black and Caspian Seas, the southern portions of Russia, and in the whole of Turkey, its numbers gradually diminishing as we proceed westward from those parts. In Germany, France, and Holland its appearance is quite accidental, and at indefinite and uncertain periods; its occurrence in the British Islands is still more seldom; indeed, the accounts of its capture here are so unsatisfactory, that we almost doubt the propriety of retaining it in our Fauna.

In its habits, manners, and general economy it closely resembles the Common Heron (*Ardea cinerea*), being an inhabitant of the marshes, and feeding upon frogs, lizards, fishes, and various aquatic insects.

It builds in trees, and lays from four to six eggs, of a bluish white.

It is now generally admitted, that notwithstanding the slight difference which exists between the present bird and its representative in America, they are really distinct. " By many of the later writers," says Mr. Selby, " *Ardea alba* has been confounded with *Ardea Egretta*, an American species, and apparently its representative in the New World. Even Temminck, whose character as a descriptive ornithologist stands deservedly in the highest estimation, has failed in pointing out the distinctive characters of the two species, and considers them in his Manual as identical, in which he has been followed by Stephens and others. Wagler, however, whose skill in detecting and accuracy in delineating specific distinctions merit the highest praise, has marked with much precision the characters of each; and Wilson, who describes the *Ardea Egretta* in his admirable Ornithology of North America, states his conviction, from a comparison of the characters of each, that the European must be a distinct species from that which he describes."

During the spring and a greater part of the summer, the adult bird is adorned with a number of beautiful, long, divided, hair-like feathers springing from the back and extending considerably beyond the tail, and which may be elevated and depressed at will. These plumes are, we believe, wholly cast off at the commencement of autumn; this peculiarity, together with its slighter form and the immaculate whiteness of its plumage, has by some naturalists been considered of sufficient importance to warrant its separation into a new genus, and although we have here retained the old name of *Ardea*, we fully concur in the propriety of this subdivision.

The whole of the plumage is of a perfectly pure white; the bill deep brown tinged with yellow about the nostrils; the space between the eyes and the orbits greyish green; irides orange; legs and feet yellowish brown.

The young are destitute of the long plumes, have the bill blackish green tinged with yellow, and the legs greenish black.

We have figured an adult male rather more than half the natural size.

LITTLE EGRET.
Ardea garzetta (Linn.)

LITTLE EGRET.

Ardea garzetta, *Linn.*

Le Héron garzette.

THE Little Egret can scarcely be considered as having a decided claim to a place in the British Fauna ; for, although one or two instances are upon record of its having been captured within the boundaries of our Island during the last half-century, still we do not know what degree of credit is to be attached to these accounts, as it is one of those birds respecting which much confusion has hitherto prevailed. The statements of its having been served up in such abundance at various feasts in the fifteenth century, recorded by the writers of that period, must be received with some degree of caution, for it is more than probable that the name was then given to another bird :—however this may be, the Little Egret is now found only in the southern por- tion of Europe, especially the countries adjacent to Asia and the Mediterranean ; a few, however, migrate periodically into France, and occasionally also into Germany ; but Sicily, Sardinia, Turkey in Europe, and the Islands of the Grecian Archipelago constitute its true habitat. Hence passing southwards and eastwards, it is abundantly spread through the temperate and warmer regions of Asia, and throughout the whole of Africa, but never occurs in the continent of America, where its place is supplied by a species closely allied, indeed, but possessing characters which sufficiently distinguish it. The young have been described by many authors as a distinct species, under the name of the " Little White Heron," because being destitute until the third year of the slender graceful plumes from the back and the occiput, it was supposed that the birds could not be identical ; this is now disproved. There is, however, a " Little White Heron " noticed by Montagu which is truly a distinct species, and the specimen he described from, the only one known to have been taken in England, is now in the British Museum.

But it is not the plumage of the young birds only that has led to confusion, for the adult birds lose their ornamental plumes after the autumn moult. Hence Buffon, who calls this bird in full feather " L'Aigrette," gives to it when unadorned the name of " La Garzette blanche."

The food of the Little Egret, like that of its congeners, consists of the reptiles and insects peculiar to the morasses among which it dwells, to which fishes and molluscous animals are also added.

Latham states, that in Egypt it is called the " Ox-keeper," from its frequenting plains where the herds of cattle are pasturing, and that it is seen " often perching on the backs of these animals to feast on the larvæ of Œstrus which infest them."

It is said to make its nest among the herbage of morasses, and to lay five white eggs.

The colour of the plumage is a pure white. In the adult birds, at least during the breeding season, the occiput is ornamented with a pendent crest of two and sometimes three long narrow feathers, and a range of slender hair-like feathers is continued down the back of the neck ; from the top of the back arise three ranges of plume-like feathers six or eight inches long, with waving shafts fine and tapering and thinly set with silky slender barbs, forming a light flowing plume ; the beak is black ; the naked skin round the eyes olive green ; irides bright yellow ; tarsi greenish black except at their lower part, where as well as on the toes the colour is greenish yellow. Length one foot eight or ten inches.

We have figured an adult in full plumage, and about two thirds of its natural size.

RUFOUS BACKED EGRET.
Ardea rufisata. (Temm.)

Drawn from Nature & on Stone by J & E Gould.

Printed by C.Hullmandel.

RUFOUS-BACKED EGRET.

Ardea russata, *Wagl.*

Le Héron roussâtre.

It is not, we believe, generally known that this little Egret has been more than once captured within the precincts of the British Isles : the first instance of its occurrence was recorded by Montagu, in the ninth volume of the Linnean Transactions, and it was afterwards more fully described in his Ornithological Dictionary under the name of Little White Heron ; and this identical specimen now forms a part of the English collection at the British Museum. It was shot in the autumn of 1805 near Kingsbridge in Devonshire, and upon dissection proved to be a female, in all probability a bird of the year, as it is destitute of the fine rufous-coloured tint with which the adults are adorned.

In Europe this species is almost entirely confined to the most southern and eastern parts, and even there it is a rare bird ; at the same time there are few species of the genus which enjoy so extensive a range, being dispersed over the greater part of Africa and Asia, and being particularly plentiful in the Himalaya and Nepaul.

Of its habits and manners we have no certain account ; but that small fish, frogs, and insects constitute its principal subsistence there can be no doubt. The specimen killed in England was observed in the same field several days among some cows and feeding upon insects.

The adult has the bill, irides, all the head and neck, and the long plumes on the back rich reddish orange ; the remainder of the plumage pure white ; the legs greenish olive ; and the nails black.

We have figured an adult of the natural size.

COMMON NIGHT HERON.
Nycticorax Europaeus. (*Steph.*)

E. Lear del et lith.

Printed by C. Hullmandel.

Genus NYCTICORAX.

GEN. CHAR. *Bill* very strong, rather longer than the head, compressed; upper mandible curved towards the point; maxilla sulcated for three fourths of its length and emarginated; culmen rounded; tomia of both mandibles straight and sharp, that of the under mandible entering within the upper one. *Nostrils* basal, longitudinal, placed in the furrow of the maxilla, and covered above by a naked membrane; lores and orbits naked. *Legs* of mean length, slender. *Toes* three before and one behind; middle toe shorter than the tarsus, exterior toe connected by a membrane to the middle one as far as the first joint. *Claws* short, falcated, that of the middle toe pectinated. *Tibiæ* naked for a short space above the tarsal joint.

COMMON NIGHT HERON.

Nycticorax Europæus, *Steph.*

Le Bihoreau à Manteau noir.

No bird, we conceive, can better show the necessity of minor subdivisions of large families than the Common Night Heron. The genus *Nycticorax* is now, we believe, universally acknowledged: seven species at least are known to us, most of which are inhabitants of remote and distant regions; one being found at Terra del Fuego, another in New South Wales, and a new one having been lately discovered in Manilla.

The Common Night Heron is the only species found in Europe, over the whole of which it is dispersed, as also over the whole of Asia, and the northern regions of Africa; and if not identical, the Night Heron of North America bears so great a resemblance to the European bird, as to require an experienced eye to detect the difference: the American birds are, however, we believe, larger in all their proportions.

Both the adults and young have been frequently killed within the British Islands, and particular instances are recorded in the works of Mr. Selby and others. It is especially abundant in Holland, France, and Germany, where it gives preference to low swampy and marshy situations, in the neighbourhood of trees and high woods. In its form the Night Heron is intermediate between the true Herons, *Ardea*, and the Bitterns, *Botaurus*, and, as may be supposed, partakes of the habits and manners of both, for although it affects more reedy and secluded situations, it nevertheless frequently resorts during the day to high trees and woods, where it may be seen perched on the topmost branches, the truth of which we can ourselves attest, having received a fine adult specimen immediately after it had been shot from a high tree in the gardens of Frogmore near Windsor: this individual evinced no fear at being approached, which enabled the keeper to make an easy prize of this rare visitor.

On the approach of evening, the Night Heron retires to the marsh or river-side, which never fails to afford it a plentiful supply of food: when fish cannot be obtained it feeds upon frogs, insects, and mice. It breeds in society much after the manner of the Common Heron; and constructs a nest, composed entirely of sticks, on the topmost branches of trees, or, when no suitable woods are near its accustomed haunts, among the reeds: the eggs are four in number, of a pale greenish blue.

Bill black inclining to yellow at the base; crown of the head, back of the neck, upper part of the back, and scapulars black with green reflections; sides of the neck, lower part of the back, rump, wings, and tail pearly grey; forehead, throat, and under parts white; from the back of the head spring three long, narrow white feathers, which are concave beneath, and lying one over the other, appear like a single plume: they can be erected at pleasure; legs and toes pale yellowish green; claws black, short, and hooked; that of the inner toe pectinated on the inner side; irides deep reddish orange; bare space round the eyes greenish blue.

The young bird during its first or nestling plumage is destitute of the plumes at the back of the head, has the culmen and point of the bill blackish brown, with the base and lower mandible yellowish green; the head and back of the neck brown, with the centre of each feather yellowish white; the front of the neck and the feathers of the breast and under surface yellowish white deeply margined with dull yellowish brown; the back and lesser wing-coverts deep brown, the centre of each feather streaked with yellowish white; greater coverts and quills deep brown, tipped with triangular spots of white; the tail brown, the legs yellowish green, and the irides bright orange. Between this state, when it is known by the name of the *Gardenian Heron*, and maturity, it acquires at each successive moulting a plumage approaching nearer to that of the adult, and in each of these stages has been described as a different species.

The Plate represents an adult and a young bird of the natural size.

COMMON BITTERN. *(Steph.)*
Botaurus stellaris. *(Steph.)*

Genus BOTAURUS, *Selby*.

GEN. CHAR. *Bill* strong, rather longer than the head; both mandibles of equal length; the upper sulcated for two thirds of its length and very gently curving from the base to the tip; tomia of both mandibles very sharp and finely serrated near the tip; lores and orbits naked. *Nostrils* basal, linear, longitudinal, placed in the furrow of the maxilla, and partly covered by a naked membrane. *Legs* of mean length. *Toes* long and slender, all unequal; middle toe of the same length as the tarsus; hind toe long, articulated with the interior toe on the same plane; claws long, subfalcate, that of the middle toe pectinated; front of the tarsus scutellated; back part reticulated. *Wings* long, rounded; the first three quills nearly equal and the longest.

COMMON BITTERN.

Botaurus stellaris, *Steph.*

Le Heron Grand Butor.

FORMERLY, when large portions of the British Islands were uncultivated, and extensive marshes and waste land afforded the Bittern abundance of retreats congenial to its habits, it was plentifully distributed over the country; but as cultivation has extended and the marshes been drained, its numbers have gradually decreased, and although not absolutely a rare bird its presence is not always to be reckoned upon, for in one year it may be tolerably common, and then for several succeeding seasons scarcely to be procured at all.

We have received specimens of the Common Bittern from Asia and Africa, but we are of opinion that Europe alone is its native habitat. At the present time it finds sufficient shelter and retreat among the marshes of Holland and other low countries, where it may fulfill the task of incubation in comparative security.

The Bittern is a solitary and shy bird, hiding itself in dense masses of reeds during the day, and seldom appearing abroad until the evening, when it resorts to ditches and the more open parts of the marshes in search of small mammalia, frogs, lizards, fishes, and various aquatic insects, retiring again to its retreat when its wants are satisfied.

No two birds can better show the necessity of subdivisions than the Bittern and the Heron, which have been until lately classed under one generic title (*Ardea*). They are equally shy and wary, yet each evinces its timidity in a strikingly opposite manner, the Heron always choosing as a place of rest, after feeding, the topmost branches of high trees, or some elevation where it can perceive the approach of danger; while the Bittern depends for security upon the covert afforded it by the thick reed-beds and other dense masses of vegetation, from whence it is not roused without considerable difficulty, and then seldom flies to any great distance. "When wounded or surprised," says Mr. Selby, "and unable to escape, it defends itself with vigour, and as it always aims at the eyes of its enemy with its strong and sharp-pointed bill, a considerable degree of caution must be used in capturing it. When attacked by a dog, it throws itself upon its back and strikes with its claws as well as with its bill; and in this manner it will keep the most resolute dog at bay, as the infliction of a stroke or two of the latter spear-pointed weapon is commonly sufficient to keep him afterwards at a respectful distance. The Bittern used to afford excellent sport in falconry; for when flown at, it immediately begins to soar, rising in spiral circles, and endeavouring to keep above its enemy. Should this manœuvre fail, it then prepares for the descent of the Hawk by setting its sharp bill perpendicularly upwards, upon which its impetuous antagonist frequently transfixes itself, or is so severely wounded as to be obliged to give up a second attack. The bellowing or booming noise of the Bittern is confined to the pairing-season, which commences in February or the beginning of March. At this time, on the approach of twilight, it rises in a spiral direction to a very great height, uttering at intervals the peculiar cry, formerly heard with superstitious dread."

In earlier times the flesh of the Bittern was esteemed a great luxury, and even now fetches a good price; it is dark coloured but not coarse, and partakes of the flavour of the hare and that of wild fowl.

The nest is composed of sticks, reeds, &c., and is generally placed near the water's edge among the thickest herbage: the eggs are four or five in number, of an uniform pale brown colour. The young are produced in about twenty-five days; they are fed by the parents until fully fledged and do not quit the nest till they are able to provide for themselves.

The sexes are alike in plumage.

Crown of the head black, glossed with bronzy green; feathers of the occiput margined with pale buff, rayed with black; from the gape a broad streak of blackish brown; all the upper surface pale buff irregularly marked with black and reddish brown, the former predominating; sides of the neck barred transversely with dark brown, the front with large longitudinal streaks of reddish brown intermingled with blackish brown; feathers of the breast blackish brown deeply margined with buff; under surface buff with narrow longitudinal streaks of brownish black; quills blackish brown, barred with reddish brown; tail reddish brown, with irregular markings of black; orbits and angles of the mouth yellow; bill yellowish green, darkest on the culmen; legs and feet pale grass green; claws pale horn colour; irides yellow.

The Plate represents a male about two thirds of the natural size.

FRECKLED BITTERN.
Botaurus lentiginosus. *(Steph.)*

AMERICAN BITTERN.

Botaurus lentiginosus, *Steph.*

Le Butor de l'Amérique.

A BIRD of this species was shot in Devonshire in the autumn of 1804; and after passing through the hands of two or three persons, who were not aware of the rarity and value of the specimen, it came into the possession of Colonel Montagu, by whom it was first described and figured in the supplement to his Ornithological Dictionary under the name of Freckled Heron, *Ardea lentiginosa*, and after whose death it was transferred with his whole collection to the British Museum. It is now ascertained that the true habitat of this species is America, and that it is only an occasional visitant to this country.

Wilson, who has described it under the specific title of *minor*, says, "This is another nocturnal species, common to all our sea and river marshes, though nowhere numerous. It rests all day among the reeds and rushes, and unless disturbed flies and feeds only during the night. When disturbed these birds rise with a hollow note, and are easily shot, as they fly heavily. Like other nocturnal birds, their sight is most acute during the evening twilight; but their hearing is at all times excellent." Wilson has also himself found and shot this species in the interior of the country near Seneca Lake, and had learned, probably from the account of Mr. Hutchins, that this bird makes its nest in swamps, laying four cinereous green eggs among the long grass. The young are said to be at first black. The stomachs of those examined by Wilson were usually filled with fish or frogs.

Dr. Richardson, in his North American Fauna, says this Bittern "is a common bird in the marshes and willow thickets of the interior of the fur countries up to the fifty-eighth parallel. Its loud booming, exactly resembling that of the Common Bittern of Europe, may be heard every summer evening, and also frequently in the day."

Top of the head dusky reddish brown; back of the neck pale yellowish brown, minutely dotted with blackish brown; a broad stripe of black on the sides of the neck, from behind the ears; upper surface dark umber brown, minutely freckled with chestnut and yellowish brown; long feathers on the shoulders broadly edged with buffy yellow; wing-coverts brownish yellow, freckled with umber brown; spurious wing, primaries, and secondaries greyish black, the tips of the latter, the lesser quills, and tail brownish orange dotted with black; chin and upper part of the throat white; front of the neck and under surface ochreous yellow with a broad stripe of mottled brown down the centre of each feather, margined on each side with a fine line of a darker tint; bill dark brown above, sides and under mandible yellow; legs greenish yellow.

We have figured an adult about two thirds of the natural size.

LITTLE BITTERN.
Botaurus minutus, (Selby.)

LITTLE BITTERN.

Botaurus minutus, *Selby.*

Le Héron blongios.

ALTHOUGH we have followed Mr. Selby in placing this bird in the genus *Botaurus*, of which the Common Bittern is the type, stil we conceive that the present species (with numerous others, possessing the same form and habits, distributed over nearly every part of the globe,) possesses characters which entitle it to form the type of a genus as distinct from *Botaurus* as that genus is from *Ardea* and *Nycticorax*. It cannot be denied, however, that it is intimately allied to the more typical *Botauri* in its solitary and secluded habits, everywhere frequenting low and swampy situations, abounding in thick coverts of reeds, willows, &c., and from which it is not driven without considerable difficulty. In England it is, and always has been, a bird of considerable rarity; nevertheless various examples have been taken at different times, so that there are few collections of any extent which do not contain one or more British specimens. On the Continent it is found in considerable abundance, especially in the southern provinces; nor is it rare in Holland and France, in both of which countries it is known to breed annually. From the seclusion of its haunts, and the difficulty of access, its nest is seldom seen : it is said to be placed in low bushes and tufts of herbage, among the thickest rushes. The eggs are five or six in number, of a pale greenish white.

The compressed form of body which so eminently characterizes the Little Bittern enables it to avoid pursuit with the utmost facility, by threading its way through the most closely compacted and intricate masses of reeds, &c., which it does with the utmost silence and rapidity. Like most other Herons, it is capable of perching; and this it often does on willows, the stems of thick reeds, &c. If forced to take wing, its flight is slow and heavy, not protracted to any great distance.

Its food consists of small fishes, frogs, snails, insects, &c.

In their adult state, the sexes offer little or no external difference in the colour of their plumage. The young are wholly destitute of the fine green of the back and top of the head, which, together with the wing-coverts, are then brown, each feather having longitudinal blotches of a darker colour. From this stage it passes through several changes of colouring, until it assumes the full plumage of maturity, which is not accomplished before the second or third moult.

Adults have the top of the head, back of the neck, whole of the upper surface, and tail glossy greenish black; middle of the wings, neck, and whole of the under surface delicate fawn yellow; bill, circle round the eye, and irides yellow; tarsi greenish yellow.

The Plate represents an adult, and a young bird in the intermediate stage, of the natural size.

WHITE STORK.

Ciconia alba. *(Bellon.)*

E Lear del. et lith.g   Printed by C Hullmandel.

Genus CICONIA.

Gen. Char. *Beak* long, straight, strong and pointed. *Nostrils* pierced longitudinally in the horny substance. *Eyes* surrounded by a naked skin. *Legs* long. *Feet* four-toed, three toes before, united by a membrane to the first joint. *Wings* moderately large, the first quill shorter than the second, and the second shorter than the third, fourth, or fifth, which are the longest.

WHITE STORK.

Ciconia alba, *Bellon.*

La Cigogne blanche.

From its familiarity, and the services which it renders to man in the destruction of reptiles and the removal of offal, the Stork has ever insured for itself an especial protection, and an exemption from the persecution which is the lot of the less favoured of the feathered tribes. Its periodical return to its accustomed summer quarters,—to its nest, the home of many generations,—has ever been regarded with feelings of pleasure ; and its intrusion within the precincts of man has not only been permitted but sanctioned with welcome. The Stork is a bird of passage, but its range is not extensive. Egypt and the northern line of Africa appear to afford it a winter retreat : with the return of summer it revisits Europe, but seldom ventures far northwards, and only occurs accidentally in the British Isles. Its scarcity in this country may perhaps be attributed rather to the drainage of our marshes, and the comparative difficulty of procuring food, than to anything uncongenial in the climate. Holland is its favourite place of residence, to which we may add the low tracts of Germany, Prussia, France, and Italy. Spain appears to be one of its winter retreats, numbers frequenting Seville at that season, " when," says Dillon, " almost every tower is peopled with them, and they return annually to the same nests." Instead of being shy and distrustful, the Stork is confiding and bold, as if aware of its privileges ; hence it may be seen on the house-tops in towns and villages, whence it wings its way to the neighbouring fields and swamps in search of food, and returns again to roost. Steeples, tall chimneys, elevated buildings, and also decayed trees, are the localities chosen for the site of its nest, a cumbrous mass of sticks and coarse materials. The eggs are generally three in number, of a pale yellowish white.

The food of the Stork consists of the various aquatic reptiles and insects which swarm in its favourite localities : its appetite is, however, somewhat indiscriminate,—snakes, mice, moles, worms, and offal being greedily devoured. With the setting in of the winter months, when the resources upon which it relies are no longer available,—when the morasses and swamps are frozen, and the ground is covered with snow,—the Storks assemble in vast multitudes, and prepare for a southward flight. Immense flocks, during the performance of this journey, are often seen in the air, passing over the country, of which instances are recorded by many writers. Like other birds of passage, it most probably begins the journey at the hour of midnight.

The males and females are alike in their plumage, having every part of a pure white, except the scapularies and wings, which are black ; the skin round the eye is also black ; the beak and feet are reddish orange.

The young have the black more inclining to dull brown.

Our figure represents an adult bird, half its natural size.

BLACK STORK.
Ciconia nigra *(Bellon)*

E. Lear del et lith.g   Printed by C. Hullmandel.

BLACK STORK.

Ciconia nigra, *Bellon.*

La Cigogne noire.

AMONG the wading birds of Europe, there are few if any which excel the Black Stork either in richness of plumage or stateliness of general aspect. Although resembling the White Stork in its habits, the present bird offers many points of difference from its well-known and familiar congener.

Instead of associating in the immediate vicinity of the habitations of man, the Black Stork is much more shy and distrustful, leading a life of seclusion among the morasses and wooded districts of the central and northern portions of Europe. The interchange of forests and tracts of marshy ground, where draining and cultivation have made but little progress, afford this bird not only food, but an unmolested asylum in which to rear its brood. Notwithstanding the length of its limbs and its semipalmated toes, it perches on trees, and builds its nest on the branches, choosing for that purpose some tall pine of ancient growth, in the depths of the forest, where its colour assimilates with the gloomy hue of the surrounding objects. It appears, however, to be a bird of migratory habits, travelling northwards and southwards with the spring and autumn. Its winter residence is not precisely ascertained, but, like all birds whose sustenance is dependent on the seasons, is doubtless in a country where the rigours of winter do not lock up the marshes and lakes with ice. Dr. Latham states it to have been met with along the Caspian Sea and at Aleppo. The preference which the Black Stork manifests for a densely wooded district is doubtless one reason why it is a bird of such rare occurrence in Holland, which in other respects is well adapted for its residence, and abounds in its favourite food, namely small fishes, frogs, worms and insects. Great Britain can scarcely lay claim to the Black Stork as one in the list of her Fauna, so few are the instances upon record of its capture in our islands.

Although shy and timid by nature, this bird soon acquires confidence and familiarity in captivity, and bears the confinement of the aviary equally well with its relative the White Stork, whose docility is proverbial.

The male and female are alike in plumage.

The head, neck, chest, and all the upper parts of the body, are blackish with purple, green, and bronze reflections ; the under surface is pure white ; the naked space surrounding the eye, that on the throat, and the beak, crimson red ; irides brown ; tarsi deep red.

The young have the beak, the naked skin round the eye and on the throat, as well as the tarsi, greenish olive, and the plumage is more inclined to reddish brown.

Our Plate represents an adult bird, half its natural size.

MAGUARI STORK.

Ciconia Maguari, (Temm.)

MAGUARI STORK.

Ciconia Maguari, *Temm.*

Le Cicogne Maguari.

WE find in this stately and fine species of Stork the first indications of a departure from the typical form of the genus *Ciconia*, and an approximation to that of *Mycteria*, which is characterized not only by a greater stoutness and solidity of the bill, but also by the recurved form of the mandibles ; a character which, in a slight degree, is perceptible in the bird before us.

In general habits and manners the Maguari Stork bears a great resemblance to its celebrated congener the White Stork (*Ciconia alba*, Bellon) ; it is not, however, a native of Europe, but obtains a place in the Fauna of this portion of the globe, from the circumstance of accidental visitors having at different times been killed in France and other places. In our islands no examples have ever been captured. America is its true habitat, the vast morasses and savannahs of that continent, both in its northern and southern portions, affording it food and shelter. Of its nidification and eggs we have no certain information.

The whole of the plumage is white, with the exception of the quills, secondaries, and upper tail-coverts, which are glossy greenish black ; beak greenish yellow at its base, passing into dull blue at the tip ; naked skin round the eye red, as is also a naked portion on the throat, which is capable of considerable dilatation ; tarsi and toes red ; nails brown ; irides greyish white.

In size, the present species is considerably larger than the White Stork, and the figure in the Plate represents the bird somewhat less than half its natural size.

SPOONBILL.
Platalea Leucorodia, (Linn.)

Drawn from life & on Stone by J & E Gould.

Printed by C. Hullmandel.

Genus PLATALEA, *Linn.*

GEN. CHAR. *Beak* elongated, strong, compressed, the point dilated and rounded, spoon-shaped, the upper *mandible* channeled round the margin, furrowed transversely at the base. *Nostrils* approaching, oblong, open and edged by membrane. *Face, head* and *chin* more or less naked. *Tarsi* long, and of considerable strength. *Toes* united by a membrane, deeply cleft, and terminating as high as the second articulation; hind-toe long, and applied entirely to the ground. *Wings*, second and third *quill-feathers* nearly equal and longest, the first rather shorter.

SPOONBILL.

Platalea leucorodia, *Linn.*

La Spatule blanche.

NATURE in her exhaustless resources exhibits great variety of means adapted to the same end, which she apparently delights to display, as if to convince us of the unlimited extent of her empire, and the powers under her command. The truth of this observation is plainly exemplified in the species under consideration. Allied to the Stork, Heron and Crane, and subsisting in a great degree on the same kind of food, still the essential modification in the structure and form of that apparatus by which it is obtained has induced naturalists to assign the Spoonbill a station as the type of a separate genus. In the Crane, the Stork and the Heron, we find the bill conical and pointed; but in the Spoonbill this organ is modelled differently; and although, as already stated, the food of the present bird nearly resembles that of the species above mentioned,—viz. mollusca, newts, frogs and their ova, together with small fishes,—it appears, if we may hazard a conjecture, a plan of construction still better adapted to its particular purpose.

Although the Spoonbill in its general manners is closely allied to the Stork, it is by no means so familiar with man, but affects localities more remote and unfrequented. In captivity, however, it acquires confidence and loses that distrust which characterizes it in freedom; and from its gentleness and inoffensive disposition, as well as from the purity of the colour of its plumage, its graceful crest, and the ease and elegance of its attitudes, adds beauty and interest to the aviary.

The genus *Platalea* thus separated is very limited, containing, as far as our researches have yet enabled us to ascertain, not more than five or six species, of which the present alone, we believe, is a native of the European portion of the globe. The British Islands, it is true, no longer afford a secure retreat for the Spoonbill, owing to the draining of our more extensive marshes and inland waters; nevertheless it still occasionally visits this country, and the author for several years past has ascertained the fact of the Spoonbill annually appearing on the coast of Norfolk, at that period of the year in which they wander in search of uninterrupted asylums; and we doubt not that this species and many others, if unmolested, would still breed with us as heretofore.

The Spoonbill is spread over Europe generally; but Holland appears to be the principal place of summer rendezvous, whence it migrates, on the approach of winter, to more southern regions, where it remains till the return of spring,—it then again retraces its course. It generally selects the tops of lofty trees for the site of its nest; sometimes, however, it chooses less elevated situations, building among rushes or reeds, and laying two or three large white eggs marked with obscure spots of red; varieties, however, are often found completely white.

The plumage of the adult bird is white, with the exception of a pale reddish-yellow band which encircles the lower part of the neck, about three inches in width on the under part, whence it gradually contracts as it extends upwards; and the crest (both of which it acquires only during the breeding season,) is of the same colour, though of a paler tint. A flesh-coloured space entirely devoid of feathers extends from the base of the lower mandible to the eye, and is continued about two inches down the throat, where it assumes a deeper yellow; beak black with a yellow tip; the irides red; the legs and feet black. Total length, two feet six inches; length of the beak varying from seven to nine inches.

The female differs from the male only in being rather smaller. The young when they leave the nest bear a general resemblance to the parent birds, with the exception however of the following particulars; viz.—the beak is not so large, softer in texture, and of a lighter colour; the shafts as well as the tips of the quill-feathers are black; the irides ash-coloured; the naked parts about the head are paler, and there is no indication of that elongated crest which ornaments the adult birds, both male and female, in the breeding season.

Our Plate represents a male in the spring plumage.

FLAMINGO.

Phœnicopterus ruber (*Linn.*)

E. Lear del et lith.   Printed by C. Hullmandel

Genus PHŒNICOPTERUS, *Linn.*

GEN. CHAR. *Beak*, thick, strong, the depth exceeding the breadth, serrated, conical towards the point, naked at the base; the upper mandible suddenly bent and curved at the tip over the inferior mandible, which is of larger size than the upper. *Nostrils* longitudinal, placed in the centre of the beak, pierced through, and covered above by a membrane. *Tarsi* of great length. *Toes* three before, united by a web as far as the nails, and one behind, which is very short and placed high on the tarsus: nails short and flat. *Wings* moderate, the first and second quill-feathers the longest.

COMMON FLAMINGO.

Phœnicopterus ruber, *Linn.*

Le Flamant.

OF all the forms in ornithology, none is more extraordinary than that of the Flamingo, whose singularly shaped bill, long and slender neck, stilt-like legs, and brilliant colouring render it a most striking object. The present form exists in all the warmer portions of the continent of Europe, Asia, Africa and America. We are acquainted with at least three or four species, but we are still uncertain whether that found in America be or be not identical with the one found in Europe; we therefore consider it best to confine our remarks to the range of the species inhabiting the Old World: we may observe, however, that should the identity of the European and American birds be hereafter satisfactorily ascertained, M. Temminck's proposed specific title of *antiquorum* must be given to the bird found in the Old World.

In Europe the only countries regularly visited by the Flamingo are those which form the coasts of the Mediterranean. It is abundant in Sicily, Calabria, and Sardinia; it occasionally occurs in France, and even in Germany, as is proved by its having been killed on the banks of the Rhine. In Asia and Africa it is very extensively spread, and, indeed, is one of the commonest birds along the whole of the African shores. Its favourite haunts in all countries are morasses, the sides of rivers, and the low muddy and sandy shores of the sea, creeks, and inlets. Admirably formed for seeking its food in these situations it is enabled from the length of its legs to wade to a considerable distance from the shore, while the corresponding length of its neck enables it to reach the bottom with its beak, which in collecting its food is placed with its upper mandible downwards, a position quite contrary to that of every other bird at present known, but for which the acute bend in the upper mandible is expressly adapted. Its food consists of small molluscous animals, the fry of fishes, and other marine productions. Though it possesses a webbed foot, this structure appears to be more for the purpose of enabling it to traverse soft and muddy places without sinking, than for the purpose of swimming, which it seldom or ever attempts. In its native haunts it is mostly seen in small bands or compa-nies, and is extremely watchful and cautious, so that it is not without the greatest difficulty a person can get within gun-shot range. Its flight when elevated in the air is rapid, and the troop assumes a wedge-shaped form, as is the case with the Wild Geese.

Not having had an opportunity ourselves of observing its nidification, we give the account published by M. Temminck and other writers, who state that it erects among the morasses an elevated mound of mud and earth, on the top of which, in a slight depression, the female deposits her eggs, placing herself astride to cover them, the great length of her limbs precluding the possibility of her assuming the usual position: the eggs are said to be two in number, of an oblong form and of a pure white.

The sexes, although differing but little in colour, may be readily distinguished by the greater size of the male. During their progress from youth to maturity, which occupies a space of four years, they undergo a considerable change of plumage; besides which, we believe, there are differences depending upon season, the beautiful rose red being characteristic of the spring and summer. The young before the first moult are of a uniform grey, with the exception of the secondaries and tail, which are black. As they approach maturity they gradually assume the snowy white and scarlet plumage of the adults, which may be thus described:

Head, neck, upper and under surface beautiful rosy white; centre of the wing bright scarlet; primaries black; bill blood red at the base and black at the tip; tarsi and toes rosy red.

The Plate represents an adult and a young bird about half the natural size.

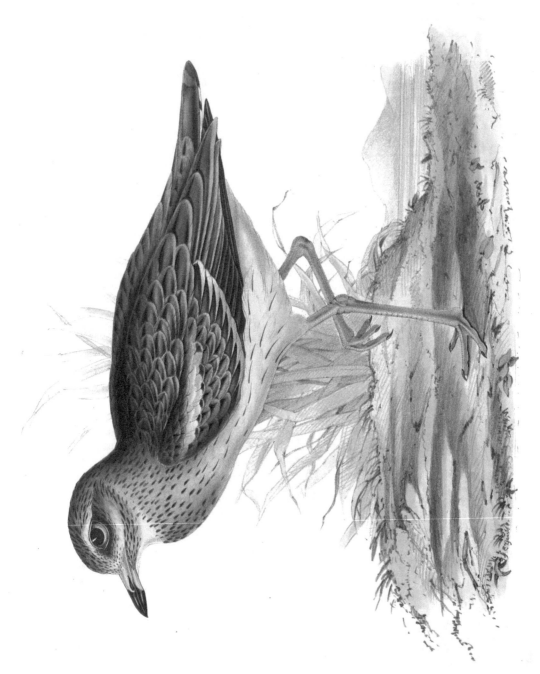

THICK-KNEED BUSTARD.
Œdicnemus crepitans. (Temm.)

Genus ŒDICNEMUS.

GEN. CHAR. *Beak* longer than the head, straight, strong, a little depressed towards the tip; *culmen* of the upper mandible elevated; lower mandible forming an angle. *Nostrils* placed in the middle of the beak, cleft longitudinally to the horny part of the beak, open before, pierced from side to side. *Tarsi* long, slender. *Toes* three before, united as far as the second articulation by a membrane which advances along their edges. *Tail* strongly graduated. *Wings* moderate, the first quill-feather a little shorter than the second, which is the longest.

THICK-KNEED BUSTARD.

Œdicnemus crepitans, *Temm.*

L'Œdicneme criard.

WE are here presented with one of those peculiar and interesting forms which serve as a link to connect two important groups. The groups to which we allude are the Bustards on the one hand, and the Plovers on the other; and we have often had occasion to remark, that while the normal or typical groups are abundant in species, the aberrant forms, which appear to be created for the purpose of filling up the intervening chasms, are restricted for the most part to a limited number of species: such is the case in the instance before us, for while the Bustards and Plovers comprise a vast multitude of species, the genus *Œdicnemus* contains at most but five or six, and these confined entirely to the regions of the Old World. The Thick-kneed Bustard is the only one of its genus which is known in Europe, in most parts of which it appears to be migratory. It arrives in the British Islands at the commencement of spring, giving the preference to elevated downs, commons, and heaths, particularly those of barren and sterile districts, confining itself, however, principally to the midland counties, being especially abundant in Norfolk, Suffolk, Kent, and Hampshire. It first appears in small companies, which soon after separate to breed. The eggs are two in number, and are placed on the bare ground, without any trace of a nest: the place of incubation is generally among loose stones and flints; and the young, which are capable of running as soon as excluded, are not to be discovered without great difficulty, their colours assimilate so closely with the surrounding objects. On the Continent it is found dispersed in similar situations, and is especially abundant, not only in the southern and eastern portions of Europe, but on the adjacent borders of Asia and Africa.

The Thick-kneed Bustard is no less distinguished for its rapidity on foot, than for its sweeping and powerful flight, which is generally performed in wide circles. Its food consists of slugs, worms, reptiles, and, not unfrequently, mice, &c.

The sexes offer little or no difference in their plumage, and the young assume the adult plumage at an early period.

The top of the head, cheeks, and whole of the upper surface brownish ash, with a tinge of vinous, each feather having a central dash of umbre brown; throat white, the same colour being obscurely indicated both above and below the eyes; a pale yellow bar passes longitudinally across the shoulders; the greater coverts are tipped with white; flanks and under surface yellowish white, the former having the shaft of each feather streaked with brown; naked skin round the eyes, the irides, and the basal half of the beak bright yellow; tarsi and toes yellow, with a slight tinge of green.

The Plate represents an adult of the natural size.

LONG LEGGED PLOVER.

Himantopus melanopterus, *(Meyer)*.

Drawn from Life & on Stone by J. & E. Gould.

Printed by C. Hullmandel.

Genus HIMANTOPUS, *Briss.*

GEN. CHAR. *Beak* long, slender, cylindrical, flattened at its base and compressed at the point; both *mandibles* channeled to the extent of half their length from the base. *Nostrils* lateral, linear. *Tarsi* very long and slender. *Toes* three before, the external and middle toes united by a membrane; *nails* small and flat. *Wings* very long, the first *quill-feather* the longest.

LONG-LEGGED PLOVER.

Himantopus melanopterus, *Meyer.*

L'Echasse à manteau noir.

THE genus *Himantopus* although widely distributed, contains, we believe, only two well-authenticated species, —the example here figured, and one very nearly allied to it from North America.

This bird, so singular in its appearance, from the extraordinary length and slenderness of its legs, has been often killed in England; but it must be classed among those birds whose visits are accidental and uncertain. It is equally scarce in Holland and the northern portion of Europe: in fact, though apparently abundant nowhere, it exhibits so wide a range, that its deficiency in point of number in any given locality is counterbalanced by its almost universal distribution. We have been presented with skins which we consider to belong to this same species, from Africa, India, the Islands of the Indian Archipelago, and, if we mistake not, from North and South America.

The Long-legged Plover, as its conformation would lead us to conclude, is a bird whose most congenial habitat is morasses, and the low flat shores of lakes, rivers and seas. Hence in the eastern portions of Europe, where it is said to arrive from Asia in small flocks, it takes up its abode along the lakes and among the vast morasses of Hungary and Russia, where, according to M. Temminck, it rears its progeny, and where it fearlessly wades in search of its food, without much chance of being carried out of its depth; but should such an occurrence happen, or the waves drift it out from the shore, it possesses, like many of the true wading birds, the power of swimming with the greatest ease and lightness; in fact, in whatever point of view we consider the Long-legged Plover, we find it adapted in the best possible manner for its habits and modes of life: few birds exceed it in the powers of flight; its wings far exceed the tail, and it passes through the air with astonishing rapidity. When on firm ground, it appears as if tottering on long and awkward stilts: but firm ground is not its congenial habitat. The egg as figured by Dr. Thienemann measures one inch nine lines in length by one inch three lines in breadth, of a pale blueish green, spotted and specked with dark brown.

In the male, the top of the head, face and under parts are white with a faint tinge of rose-colour; back of the head and neck black; back and wings black with green reflections; tail light grey; beak black; irides crimson; tarsi and feet fine orange-red. In some stages of plumage the head and neck are perfectly white, the black being, in all probability, the plumage of summer.

The female differs from the male in having the back brown instead of black, with green reflections.

The young have their colours altogether more obscure, with a brown tinge; the tarsi and irides are also less brilliant.

We have figured a male in perfect plumage, nearly of the natural size.

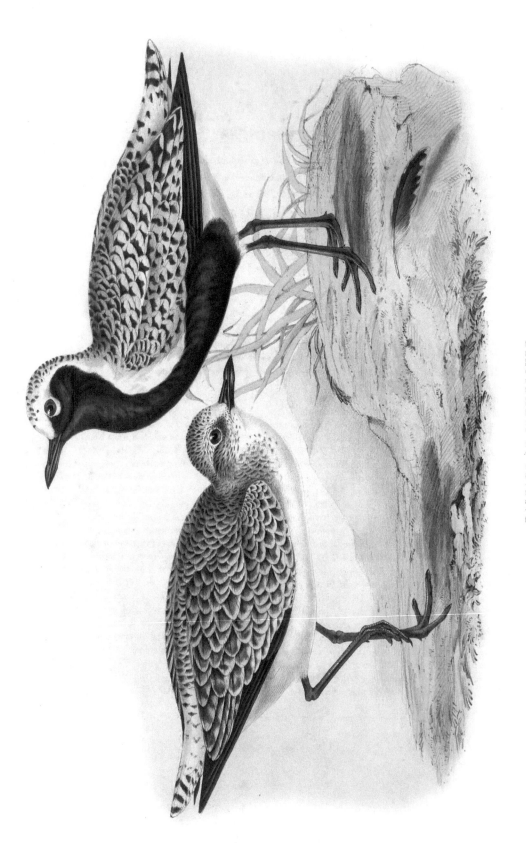

BASTARD OR GREY PLOVER.
Squatarola cinerea. *(Cuv.)*

Drawn from Nature & on Stone by J. & E. Gould.

Printed by C. Hullmandel.

Genus SQUATAROLA.

GEN. CHAR. *Bill* rather strong, cylindrical, straight, nearly as long as the head; the tip, or horny part, about half the length of the whole bill, tumid, and arched, with the *tomia* bending inwards. *Nasal groove* wide, half the length of the bill. *Mesorhinium* depressed below the level of the tip. *Nostrils* longitudinally pierced in the membrane of the groove, linear, oblong. *Wings* rather long, acuminate, with the first quill-feather the longest. *Legs* slender, of mean length, naked above the tarsal joint. *Feet* four-toed, three before and one behind; front toes joined at their base by a membrane, that portion of it between the outer and middle toe being the longest. Hind toe very small or rudimental. *Tarsi* reticulated. *Plumage* thick, close, and adpressed.

GREY PLOVER.

Squatarola cinerea, *Cuv.*

Le Vanneau Pluvier.

THE Grey Plover is the only European example of the genus *Squatarola*, a genus of more than ordinary interest to the ornithologist, possessing as it does characters which seem to place it in an intermediate situation between the genus *Charadrius* on the one hand and that of *Vanellus* on the other. In its most striking peculiarities, both as regards its general form and the nature of its periodical changes of plumage, it exhibits a striking affinity to the well-known Golden Plover; in fact, in some stages of its plumage, it requires a nice attention to other points to distinguish it from that bird: for example, during the first autumn and winter of its existence, each feather has its edges spotted and margined with yellow, as the single feather in the foreground of our Plate will illustrate. The next change consists in the loss of the yellow colour, which is exchanged for grey, a style of plumage which ever after characterizes this bird, excepting in the commencement of the breeding season, and during the subsequent moult, when the greater part of the under surface is, as in the Golden Plover, of a uniform rich and glossy black. Trusting to these characteristics alone, we should be ready to assign to the bird a place in the genus *Charadrius*; but on examining the feet we should immediately discover our mistake, for although we do not find a well-developed hind toe, still we are presented with one in a rudimentary condition, indicating the fact of its departing from the typical form of the genus *Charadrius*, and its alliance to that of *Vanellus*, between which it becomes a connecting link: hence we agree with Baron Cuvier in the propriety of constituting the genus *Squatarola* for its reception.

The range of the Grey Plover is very extensive. It is found in the northern portions of the continents of Europe and America, everywhere preferring the borders of the sea and the mouths of large rivers, particularly where low, flat, muddy shores extend, abounding with food expressly adapted to it, such as worms, various kinds of insects and their larvæ. At some seasons it is very abundant on our shores; at others more sparingly diffused, but never, we believe, altogether absent: they appear in the greatest abundance while performing their periodical migrations, in the months of April and May, when their numbers are greatly augmented.

Along the coasts of Lincolnshire, Norfolk, Suffolk, Essex, and Kent, at such seasons they appear in flocks, some individuals having the breast wholly black, others mottled with black and white, while some few have not yet begun to exhibit this change; which varied appearances depend on the maturity of the bird, and on the degree of forwardness attained for the great purpose of reproduction.

But little can be correctly affirmed regarding its nidification; and it is still questionable whether it is to be deemed a bird which regularly breeds in our island. M. Temminck informs us that it is common in the regions of the arctic circle and on the confines of Asia. The eggs are four in number, of a light olive blotched with black.

The Grey Plover, like its relation the Golden Plover, is an active bird, running rapidly along the flat shores with his head depressed and his body in a horizontal position; nor is it less remarkable for its powers of flight.

In consequence of the remarks on the plumage above given, it will only be requisite for us to describe the livery which characterizes the bird during the first autumn. The whole of the upper surface, together with the sides of the chest, are beautifully bespangled with yellow and brown on a dark olive-grey ground; the under parts white; irides, beak, feet and legs, blackish olive.

The Plate represents two adult birds: one in its spring, the other in its winter plumage, both of the natural size; and it will be observed that the black chest of the bird in its spring plumage is bordered by a band of pure white.

LAPWING.

Vanellus cristatus, (Meyer).

Genus VANELLUS.

Gen. Char. *Bill* shorter than the head, straight, slightly compressed; the points of both mandibles horny and hard. *Nasal groove* wide, and reaching as far as the horny tip. *Nostrils* basal, linear, pierced in the membrane of the nasal groove. *Legs* slender, with the lower part of the tibiæ naked. *Feet* four-toed; three before, and one behind, united at the base by a membrane; hind toe very short, articulated upon the tarsus. *Tarsi* reticulated. *Wings* ample, tuberculated or spurred; the first three quill-feathers notched or suddenly narrowed towards their tips, and shorter than the fourth and fifth, which are the longest.

LAPWING.

Vanellus cristatus, *Meyer.*

Le Vanneau Huppé.

The Lapwing, or Peewit, both with respect to the elegance of its plumage and the general outline of its contour, and the light and graceful manner in which it trips along the ground, together with its habits and economy, is one of the most interesting of our native Plovers. When we consider that all this grace and elegance is appointed by nature to add life and cheerfulness to bleak moorlands and swampy plains, far from the habitual resort of man, we cannot but feel that those desolate spots in creation are not without their peculiar attractions. Those of our readers who would wish to observe this bird in a state of nature will be amply repaid by a visit to its native districts; and as there are no heaths, wide moorlands, or swampy places of any extent throughout the British Islands without the presence of the Peewit during summer, their natural habits and manners may be investigated with great ease and with but little expense or trouble.

This species is distributed over the whole of Europe, everywhere inhabiting situations similar to those it occupies in our own island. We have also seen specimens in collections from India and Africa. Although this individual species is confined to the Old World, America is not without this peculiar form, as the collection formed by Capt. P. P. King during his late survey of part of the southern coast, and since presented by him to the Zoological Society, fully confirms.

The flight of the Lapwing is heavy, flapping, and apparently performed with considerable exertion. If the breeding-place be approached, the male utters his call of *Peewit* repeatedly, and at the same time exhibits a series of aërial evolutions peculiar to himself; and although no great velocity distinguishes his flight, his numerous turns are made with great quickness and dexterity.

Its nidification, like that of the rest of its allied race, takes place on the bare earth, no nest being made for the reception of the eggs, which are four in number, of an olive colour blotched and spotted with black.

They perform the task of incubation at an early season of the year, commencing on our heathy moorlands in the month of March, soon after which they deposit their eggs, which are eagerly sought for as a luxury for the table; hence immense numbers are annually gathered and transmitted to the various markets: nor are the birds themselves less esteemed. Their food consists of insects, worms, and slugs.

The sexes are much alike in plumage; the male, however, exhibits a richer tone of colouring and a longer occipital crest. Some little variation takes place between the summer and winter dress, the male in the former season being adorned with a fine black throat, which changes to white in winter. The young attain the adult plumage in their second year.

Adults have the top of the head, crest, and breast black; the upper surface black, with green and bronzed reflections; many of the back feathers edged with brown; the under parts pure white; the tail-feathers white largely tipped with black; the outer tail-feathers white; the upper and under tail-coverts rufous; the beak black; and the legs reddish brown inclining to purple.

The Plate represents a male and female in the summer and winter plumage.

KEPTUSCHKA LAPWING.
Vanellus keptuschka. (Temm.)

Drawn from Nature & on Stone by J & E. Gould.

Printed by C. Hullmandel.

KEPTUSCHKA LAPWING.

Vanellus Keptuschka, *Temm.*

Le Vanneau Keptuschka.

It affords us considerable pleasure that we are enabled to illustrate the old and young of this very interesting and rare species of *Vanellus*, from a fine adult male kindly forwarded to us by M. Lichtenstein of Berlin, and from a young bird in the collection of the Zoological Society of London, to whom it was presented by their valued correspondent Keith E. Abbott, Esq., of Trebizond. We are also indebted to M. Temminck for the use of a spirited oil painting of this bird, taken from an individual killed in France. M. Lichtenstein considers the *Vanellus Keptuschka* and the *Vanellus gregarius* to be one and the same species, and we find on reference to the 'Systema Avium,' that this was also the opinion of the lamented Dr. Wagler, who in his monograph of the group has given the preference to the specific appellation of *Keptuschka*.

The eastern provinces of Europe constitute the only portions of our quarter of the globe inhabited by this species: it also frequents the marshes of Siberia, and is common in Persia and Asia Minor.

Although closely allied to the Common Lapwing of our island, this bird and a few other species may hereafter be considered sufficiently distinct from the typical form of the genus to constitute a separate group, particularly when we have acquired a knowledge of their habits, mode of flight, &c.

The adult male has the forehead and a broad stripe surrounding the crown pale buffy white; crown of the head, space between the bill and the eye, and a narrow line behind the eye deep black; throat and sides of the neck buff, which is palest on the chin; back of the neck, back, rump, scapularies, and wing-coverts light brownish grey tinged with olive; secondaries pure white; quills deep black; breast dark brownish ash, gradually passing into deep black on the abdomen, which latter colour terminates posteriorly in rich chestnut; thighs, vent, under tail-coverts and two outer tail-feathers pure white; the remainder of the tail-feathers pure white, with a more or less extensive mark of deep black near their extremities, presenting the appearance when the feathers are closed of one large irregular patch; bill and feet black.

The young has the feathers of the crown dark brown in the centre, margined with buff; only a faint indication of the band surrounding the crown; sides and back of the neck, breast, upper surface, and wings dull olive brown with paler margins; chin and abdomen white; secondaries, quills, and tail as in the male.

The figures are of the natural size.

SPUR-WINGED PLOVER.
Pluvianus spinosus.

Drawn from Nature & on stone by J.&E.Gould.

Printed by C.Hullmandel.

SPUR-WINGED PLOVER.

Pluvianus spinosus.

Charadrius spinosus, *Auct.*

Le Pluvier armé.

THIS species of Plover is said occasionally to visit the southern and eastern portions of Europe; and when we consider the wide range it possesses in the adjoining countries, we cannot wonder at this circumstance, it being abundant over nearly the whole of Northern Africa, Asia Minor, &c. Dr. Latham informs us that it " inhabits Russia, and is frequent near Aleppo, about the river Coic. The Spur-winged Plovers are very numerous and exceedingly noisy; have a hasty and almost continual movement of the head and neck, drawing them up briskly, and then stretching them quickly forward, almost as if they were making hasty and eager bows."

Of its nidification nothing is at present known.

The sexes are so closely alike in plumage that one description will serve for both.

Bill, crown and back of the head, a broad stripe down the centre of the throat, breast, abdomen, primaries, and the tips of the tail-feathers deep black; the outer feather of the latter finely tipped with white; sides and back of the neck, under surface of the wings, secondaries, greater wing-coverts, flanks, vent, upper and under tail-coverts pure white; back and remainder of the wings greyish brown tinged with olive; legs, feet, and spur on the shoulder brownish black.

Our Plate represents a male of the natural size.

GOLDEN PLOVER.

Charadrius pluvialis, *(Linn).*

Genus CHARADRIUS.

GEN. CHAR. *Beak* slender, straight, compressed, shorter than the head ; *nasal furrow*
prolonged more than two thirds ; *mandibles* enlarged towards the tip. *Nostrils* basal,
jagged, cleft longitudinally in the middle of a large membrane which covers the nasal
fossa. *Legs* moderate, or long and slender. *Toes* three, directed forwards, the external
united to the middle one by a short membrane ; the inner toe free. *Tail* square, or
slightly rounded. *Wings* moderate ; first *quill-feather* the longest.

GOLDEN PLOVER.

Charadrius pluvialis, *Linn.*

Le Pluvier doré.

THE Golden Plover is extensively spread over the whole of the northern portions of Europe, and is by no
means uncommon in our Islands, inhabiting heaths, downs, and swampy moors during the summer, but con-
gregating near the coast and about the marshy inlets of the sea in autumn and winter. Its breeding place,
however, will generally be found upon the heath-covered hills of our northern counties, and the Highlands of
Scotland, the female depositing her eggs on the ground, four in number, of a large size, exceeding those of
the Lapwing, of a dull olive-coloured ground blotched with black.

The habits of the Golden Plover, as well as those of the rest of its congeners, may be denominated exclusively
terrestrial ; for although distinguished by a rapid and sweeping flight, its characters, structure and powers, are
such as to qualify it in an especial manner for running on the bare turf and among the heath of the mountains,
which it does with great quickness and agility ; and as nature perfects those endowments the earliest, on
which depend the means of maintaining existence or avoiding foes, so we see the young, just excluded from
the egg, covered with dusky brown, crouching or running with great celerity, and yet incapable of flight,—a
power which they do not possess until after a considerable period.

The difference that exists in the plumage of the Golden Plover at particular seasons of the year is sufficient,
without a knowledge of the change, to produce a deception as to the identity of the species. In winter, the
general colour of the upper surface is dusky with numerous spots of yellow, lighter beneath ; but in March, a
few black feathers appear on the breast and under parts, which are augmented in number during the succeeding
month ; and in May a broad expanse of jet black, beginning above the beak, and passing over the cheeks and
sides of the neck, covers the throat, breast, and under parts of the body. The margins of this black close
abruptly on a line of white, which continues its course from the forehead along the neck and sides, gradually
blending with the rest of the plumage. As soon as the season of incubation is over, the black feathers disappear,
as well as the white marginal line, the dusky plumage of the winter returning.

The cause of this change in plumage is not clearly understood ; it is, however, produced by a partial moult
which takes place in the spring and autumn ; so that the black colour is not superinduced upon the old feathers,
but is the original colour of the new ones ; nor does this tint fade in these feathers so as to become grey,
but the feathers themselves gradually fall off in the autumn, the grey ones succeeding. The whole plumage,
with the exception of the primaries, which are moulted but once in the year, is subjected to a similar law :—
we do not here mean to say that in all birds which undergo periodical changes in the colour of their plumage
this double moult takes place ; but in the present instance we have ascertained such to be the case. The
young of the year differ but little from the adult in winter, with the exception of a yellower tint pervading
the whole body.

The bill is dusky ; eyes dark hazel ; head and all the superior parts of the body dark brown, beautifully
spangled with golden yellow ; legs and feet olive-brown. Length ten inches : weight seven or eight ounces.

Its flesh is extremely delicate, and is much sought after for the table ; hence there are multitudes annually
brought to the London markets for sale.

We have represented the adult bird in its summer and winter dress, the black breast being the characteristic of the former :—the sexes are not distinguishable by their plumage.

DOTTRELL.

Charadrius Morinellus. /Linn/

Drawn from Life & on Stone by J & E Gould.

Printed by C. Hullmandel.

DOTTRELL.

Charadrius morinellus, *Linn.*

Le Pluvier guignard.

THE natural history of this species is less perfectly known than that of many others which are much more rare. It is seen in several parts of England, and in considerable numbers, but only at two periods during each year, in its passage to and from that country in which it breeds; yet where that very important part of its economy is accomplished to any extent, has been but partially proved.

These birds make their first appearance every year in the month of May, sometimes as early as April, and are then in their finest plumage. The female frequently weighs upwards of four ounces, and measures almost ten inches in length; the male weighs only three ounces and a half, and measures but nine inches and a half. The plumage of the sexes is not very dissimilar; and it has happened to us, that the largest in size, as well as the finest in plumage we have ever been able to procure, have invariably, on internal examination, proved to be females. The beak is dusky; irides hazel; the forehead speckled with brown and white; crown of the head much darker, the middle of each feather being nearly black and edged with light brown; from the beak, and passing over the eye on each side, is a broad band of white, which extending backwards almost unite at the nape of the neck; chin and throat white, with small elongated brown spots; the whole of the neck below ash-grey; back and wing-coverts light yellowish brown, each feather edged with pale fawn-colour; lower part of the neck white, occasionally bounded above with a narrow line of black; breast rich orange; abdomen black; region of the vent and under tail-coverts greyish white; quill-feathers dusky brown; the tail-feathers olive brown,—both margined with pale ferruginous; legs dingy yellow brown; toes darker. Young birds of the year have the crown of the head mottled with brown and white, the white mark over the eye less conspicuous, the colours on the upper parts more dull, with the whole under surface of the body pale ferruginous and dusky.

It is stated of these birds, that they are more abundant in Asia than in Europe; rather common during winter in the Grecian Archipelago and the Levant; are seen, on their passage, in Germany and France, but very rarely in Holland. They visit Sweden, Dalecarlia, and the Lapland Alps, and breed in the northern parts of Russia and Siberia.

In our own country they frequent the downs of Sussex, Hampshire, Wiltshire, Berkshire and Cambridgeshire, resorting to the open fallow-grounds in their vicinity for food, which consists principally of insects and worms. They first appear about May, in small flocks, or *trips* as they are called, of from four or five to ten birds each, on their passage northwards, and return at the end of August recruited in numbers by the addition of their offspring, and we have at that season seen twenty and sometimes thirty together. We have learned also from old shepherds on the Royston and Cambridge hills, that these birds were formerly much more numerous there than they are at present. They are also seen in Lincolnshire, Derbyshire and Yorkshire. They appear in the neighbourhood of Carlisle in May, remaining ten days or a fortnight, and then depart for Skiddaw and the adjoining mountains, where they are said to breed annually. On Skiddaw in particular, a few of these birds have been seen and shot in the month of June, and their nests and eggs taken.

Montague, and also Colonel Thornton, saw an occasional pair of Dottrell in Scotland, at a season which warranted them in concluding that some of these birds produced their young in that country. We do not remember any record of their having been seen in the Hebrides, nor does Mr. Low include them in his Fauna of Orkney and Shetland.

Dr. Latham, in his General History of Birds, informs us, that in the district of Aberdeenshire, called Braemor, (being the most elevated part of the country,) these birds hatch their young on dry mossy ground near to, and on the very summits of, the highest parts; sometimes in the little tufts of short heather, or moss, which are to be found in those elevated grounds: even in so exposed a situation they take so little trouble to form their nest, that were it not by the eggs, no person could suppose there was one. The hen sits three weeks, and the young birds make their appearance about the middle of July: they rarely lay above three eggs, and generally bring forward as many young.

The eggs of these birds are so difficult to obtain, that we only know one collector who possesses them. They are one inch eight lines long, by one inch two lines and a half in breadth, light olive brown, blotched and spotted with black:—these specimens were procured from the Grampian Hills.

About the periods of their passage to and from their breeding-ground, as before referred to, a few of the Dottrell are to be seen occasionally in the London markets, and always command a considerable price for so small a bird, usually selling readily at six shillings per couple. A young bird in good condition is said to be of exquisite flavour.

RING DOTTREL.
Charadrius hiaticula. (Linn.)

RING DOTTRELL.

Charadrius hiaticula, *Linn.*

Le Grand Pluvier à collier.

Of all those elegant birds which constitute the group of the genus *Charadrius* generally termed Dottrells, the present beautiful species is by far the most common and universally spread, being an inhabitant, not only of our own shores, but equally so of those of temperate Europe in general, as well as of North America. Its favourite localities are flats along the shore, particularly where the sea at its ebb retires to a distance, and leaves extensive beds of sand or shingles. It is not, however, confined entirely to the margin of the sea, being not unfrequently met with on the borders of lakes and at some distance from the mouths of large rivers. It appears to be a bird which makes a permanent residence in our own and similar latitudes, breeding along the shore among broken shells and gravel, merely hollowing out a small depression for the reception of its eggs, which are of considerable size in proportion to the bulk of the bird,—a circumstance which may be noticed as general among the *Charadriadæ.* Their colour is yellowish white, streaked and dotted with irregular marks of black, which are most numerous at the larger end.

The young, while yet covered with down, and hatched a few days only, run with great quickness and dexterity, separating and squatting for concealment among the loose stones and long weeds of the shore on the appearance of danger, while the parents exert every artifice to draw off the intruder to a distance, uttering their mournful cry, and feigning inability to escape pursuit.

The food of the Ring Dottrell, which it searches for on the shore, consists of worms and insects of various species.

The difference between the summer and winter plumage consists rather in the depth of the black and pureness of the white, during the season of incubation, than in any decided change. In winter, in fact, the plumage approaches to that of the immature bird of the year, in which the black band on the chest is indicated only by a dark tint of brownish grey.

The adult colouring of the male in summer consists of a brownish grey over the whole of the upper surface, the head being white with a black band stretching over the top from eye to eye, and a band of a similar colour passing from the forehead, at the base of the upper mandible, beneath the eye and over the ear-coverts ; a broad black band occupies the chest, becoming narrower at the back of the neck, where the extremities meet ; beak bright orange colour merging into black at the tip ; naked circle round the eyes and tarsi orange.

The female differs in having the coronal band and that of the chest of smaller size and of a somewhat browner hue.

Besides having only indications of the pectoral band, the immature birds have the feathers of the upper surface of a light brown, edged with yellowish, the coronal band wanting or very obscure ; the beak blackish ; and the tarsi dull yellow.

The Plate represents a male and female in the spring plumage, of the natural size.

Drawn from Nature & on stone by J & E Gould.

LITTLE RING-DOTTREL.
Charadrius minor. *(Meyer)*

Printed by C. Hullmandel.

LITTLE RING-DOTTRELL.

Charadrius minor, *Meyer*.

Le Petit Pluvier à collier.

WE are indebted to our friend Mr. Henry Doubleday, of Epping, for the loan of an example of this elegant little Plover, which he informs us was taken at Shoreham in Sussex. From the extreme youth of the specimen transmitted to us, it is clear that it must have been bred on the spot; and it is worthy of notice that the person who killed it affirms that he has long suspected the present bird to be a resident on that part of the coast, from having remarked that he could always perceive a difference in the note of this bird from that of either of the other species. Whether this Plover habitually resorts to our shores or not, it may now reasonably claim a place in the Fauna of our island; and we are glad of the opportunity of introducing it to the notice of British ornithologists, and still more so that the only British-killed specimen should have fallen into the hands of an individual so zealous in the collection of our native birds as the gentleman above mentioned. On the Continent it is by no means a scarce bird; we learn from the *Manuel* of M. Temminck that it is abundant in the South of Germany as far as Italy, and that it is occasionally found as a bird of passage in Holland, ever giving the preference to the borders of large rivers rather than the shores of the sea. We have compared it with American specimens, and can attest that they are specifically different.

Its general habits, manners, and mode of life are strictly in accordance with the Common Ring-Dottrell; like that species it constructs its nest on the sand and shingles which border the water's edge. The eggs are four or five in number, of a yellowish white colour, marked with blotches of black and brown.

The adults of both sexes are nearly alike in plumage; the young, on the contrary, do not acquire the collar and black markings until the second year. From the Common Ring-Dottrell, the only bird in Europe with which it could be confounded, it differs in being much smaller in size, in having the beak entirely black and comparatively small, and in the fleshy colouring of the tarsi.

The adults have the bill black, a band of the same colour passing from the bill to the eye, and extending over the ear-coverts; the forehead pure white, above which on the crown a black band passes from eye to eye; the occiput grey, beneath which a white circle spreads from the throat round the neck; this is succeeded by a black band, broad on the chest, but narrowing until it meets at the back of the neck; the whole of the upper plumage, with the exception of the rump, which is white, of a fine brownish grey; under surface white; feet and legs flesh colour; irides hazel.

The young entirely want the black collar and facial markings, the crown of the head and face being brownish grey; in every other respect they resemble the adults, except that a brownish tint pervades the whole of the upper plumage and that every feather is edged with a lighter margin.

The Plate represents an adult, and a young bird of the first autumn, of the natural size.

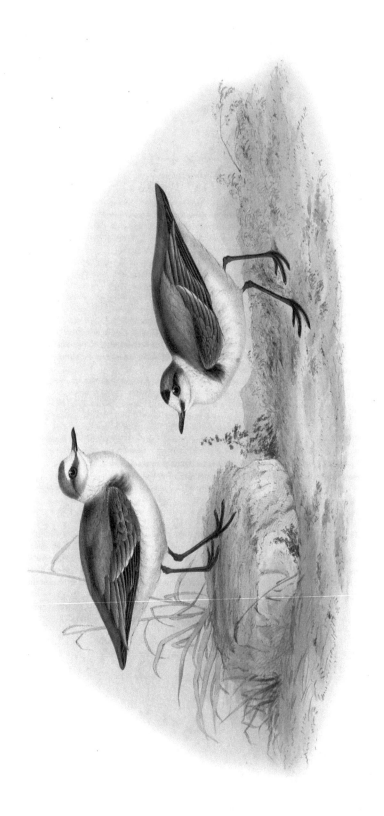

KENTISH PLOVER.

Charadrius cantianus, /lath/

KENTISH PLOVER.

Charadrius Cantianus, *Lath.*

Le Pluvier à collier interrompu.

In its habits and manners, as well as in localities, this interesting little Plover is closely allied to the Ring Dottrel, *Charadrius hiaticula*; and we have every reason to suppose that it is often mistaken by casual observers for that bird, as when seen in a state of nature, unless approached very nearly, they are scarcely distinguishable : still no Ornithologist could mistake it upon actual examination; its smaller size, black legs, and the rufous mark behind the head which characterizes the male, affording a marked ground of distinction. As the Kentish Plover is a species still in great request by most collectors of our native birds, we have the pleasure of stating, that if sought for in the localities hereafter mentioned it will assuredly be met with, our own experience enabling us to assert that it annually breeds in many parts of this Island. We have received it in considerable abundance from Great Yarmouth in Norfolk, which, as far as we have been able to ascertain, forms its northern boundary in this country. It is also found along the flat and shingly beaches of Kent and Sussex; and we may particularize Selsey beach, the immediate neighbourhood of Hastings, and Shellness near Sandwich, as places where, if sought for in the months of May, June and July, it is sure to be met with. During the last season several pairs were shot near Sandwich by the Rev. George Clayton, of Much Hadham, near Bishops Stortford, Herts. The opposite coasts of France and Holland, where a similar character of shore prevails, are also among the places to which it pays annual visits.

As we have before mentioned, the actions of the Kentish Plover are strictly similar to those of the Ring Dottrel, in whose company it is often found, the instinct of both species leading them to prefer the sea-shore, its inlets and creeks, seldom venturing from these localities to visit the fresh waters. Like the rest of its genus, it is quick and active in its motions, running with great celerity, with its head depressed below the level of its back, over the loose shingles and muddy flats of the shore, and occasionally taking short and circular flights, returning almost to the same spot, uttering while on the wing a brief and mournful note, repeated at short intervals.

Its food consists of small marine insects and worms, which it searches for among the loose stones, sand, and oozy mud; to which it also adds the smaller kinds of slender-shelled bivalves.

It lays its eggs, which are five in number and of a yellow olive marked with irregular spots and blotches of dark brown, in depressions on the naked sand, or among the shells and shingles of the beach.

M. Temminck informs us that the moult of this bird is single, taking place in autumn, as he has often had opportunities of observing.

In the adult male the colours of the plumage are as follows. Forehead, eyelids, a band on the back of the neck, and all the under parts, of pure white; the space between the eye and the beak, a band on the forehead, and a large spot on each side of the breast, of a deep black; behind the eye, a large dark grey mark; the head and back of the neck, of a light reddish brown; the upper parts, of a brownish ash colour; the quill-feathers slightly edged with white; the two lateral tail-feathers white, the third whitish, and the others brown; beak, irides and feet, black.

The female wants the black band on the forehead, its place being occupied by a little transverse bar; the sides of the breast, the space between the eye and the beak, and the region behind the eyes, are ashy brown; and the back of the head and neck is tinged with grey.

The young of the year differ from the adult females chiefly in having the feathers of the upper part of a light ashy brown, each feather being edged with a lighter tint of the same colour; and the lateral markings of the breast being indicated by light brown.

We have figured a male and female, in their adult plumage, of the natural size.

RED-CHESTED DOTTREL.
Charadrius pyrrhothorax. *(Temm.)*

Drawn from Nature & on stone by J. & E. Gould.

Printed by C. Hullmandel.

RED-CHESTED DOTTEREL.

Charadrius pyrrhothorax, *Temm.*

An example of this bird has been forwarded to us by our valued friend and correspondent M. Temminck as a species inhabiting Europe, and of which, he informs us, a description will appear in the forthcoming part of his ' Manuel.'   No information having been transmitted with the specimen, we must content ourselves with giving a faithful representation of it, and with stating that as we have seen it in many collections from India, we are consequently disposed to consider it as only an occasional visitant to the eastern portions of the European continent.

A broad stripe crossing the forehead, passing under the eye, over the ear-coverts, and down the sides of the neck blackish brown; crown of the head, back of the neck, all the upper surface and tail greyish brown; primaries blackish brown; throat, abdomen, and under tail-coverts white; breast rich reddish chestnut, which colour gradually blends with the white of the throat and abdomen; bill, legs, and feet black.

The young bird has the forehead, a faint line over the eye, and all the under surface pure white; stripe between the bill and eye, and the ear-coverts pale brown; all the upper surface greyish brown, but of a much lighter tint than in the male; bill, legs, and feet black.

We have figured an adult, and a young bird of the natural size.

OYSTER CATCHER.
Hæmatopus Ostralegus. (Linn.)

Drawn from Nature & on Stone by J. & E. Gould.

Printed by C. Hullmandel.

Genus HÆMATOPUS.

GEN. CHAR. *Bill* longer than the head, straight, strong, the point much compressed, forming a wedge; culmen of the anterior part slightly convex; upper mandible with a broad lateral groove extending one half the length of the bill; mandibles nearly equal and having their tips truncated. *Nostrils* basal, lateral, linear, pierced in the membrane of the mandibular groove. *Legs* of mean length, naked for a short space above the tarsal joint. *Tarsi* strong. *Feet* three-toed; all the toes directed forward, and united at their base by a membrane. *Nails* strong, broad, slightly falcate, and semi-acute. *Wings* of mean length, with the first quill-feather the longest.

OYSTER-CATCHER.

Hæmatopus ostralegus, *Linn.*

La Huiterier pie.

THE only species of this widely diffused but restricted genus which can claim to be considered as European is the bird before us, and we have every reason to regard it as indigenous, not only in the British Isles, but throughout the whole of the Continent. The localities to which it gives preference are the low muddy shores bordering the sea, salt marshes, and inland saline lakes. The whole of its actions are characterized by considerable liveliness and spirit : it runs along the level sands with great swiftness, nor is it less distinguished for its sweeping velocity when on the wing ; in addition to which it swims with ease and address, although it does not habitually take to the water : indeed, it is only when wading far from the shore, and finding itself out of its depth, that it resorts to the expedient of swimming to effect its return. Its robust and powerful frame admirably adapts it for the efficient use of its strong hard bill in obtaining its prey, such as limpets (which require considerable force to detach them from the stones), bivalves, crustacea, and marine worms. As its name implies, it is said to be extremely dexterous in opening the shells of oysters in order to obtain the animal within, which is known to be a favourite article of its food.

In winter the Oyster-catcher is gregarious, assembling together in considerable flocks, which separate on the approach of spring, when each pair retires to its peculiar breeding-station.

It is very common in Lincolnshire, Norfolk, and all parts of our coast where a marshy stripe of land borders a long sandy beach, among the shingles of which it deposits its eggs, which are four in number, of a light olive colour blotched and otherwise marked with black. While the female is engaged in the task of incubation, the male keeps assiduous watch, and gives notice of the approach of danger by a sharp and peculiar kind of whistling cry. The young quit the nest on the day of their exclusion from the egg, and are assiduously attended by the parents, which continually sweep round any intruder, and assail him with loud cries. The young attain at an early age the adult livery, without undergoing any intermediate gradations of plumage.

The sexes are alike in their outward appearance, and the only difference in their summer and winter dress consists in the presence of a white crescent-shaped mark half round the throat during the latter season.

The bill is reddish orange at the base, becoming lighter at the tip ; the legs orange red ; irides crimson ; the whole of the plumage is black, with the exception of the rump, a band across the base of the quills, and the under surface, which are pure white.

The Plate represents an adult male in the summer plumage, of the natural size.

GLOSSY IBIS.

Genus IBIS.

GEN. CHAR.　*Beak* lengthened, slender, arched, large at its base, depressed, obtuse, and rounded at the point; the upper mandible grooved throughout its whole length.　*Nostrils* near the base on the upper part of the beak, oblong, narrow, encircled by a membrane.　*Face* naked, as is often also a part of the head and neck.　*Legs* moderate and slender, naked above the knee.　*Toes* three before and one behind; the former being webbed as far as the first joint; the hind toe long, and resting closely on the ground.　*Wings* moderate, the first quill-feather shorter than the second and third, which are the longest.　*Tail* short and square.

GLOSSY IBIS.

Ibis Falcinellus, *Temm.*

L'Ibis Falcinelle.

WHILE every temperate and tropical country of the globe possesses various examples of this widely spread genus, the present species is the only one, as far as is known, which passes over the border line of Africa and Asia, and takes up its residence in the Southern and temperate countries of Europe. The Glossy Ibis is tolerably abundant in all the swampy and marshy districts of its south-eastern portions, particularly Hungary, Turkey, and the Archipelago: it passes hence, but in much less abundance and at uncertain intervals, to the more central and western portions; and occurs, though very rarely, in Holland and the British Islands. Along the course of the Nile and in the adjacent provinces of Africa, this handsome bird appears to find a most congenial situation, and is as abundant there at the present day as it was in ancient times, when it was regarded as sacred, and embalmed equally with the *Ibis religiosus*, or Abouhannes of Bruce.

As is the case with the other birds of its tribe, the food of the Glossy Ibis consists of worms, slugs, lizards, freshwater mollusca, and aquatic vegetables.

The graceful proportions of this bird, the elegance of its actions, together with the resplendent lustre of its plumage, render it one of the most interesting of the Waders, and we have to regret that our knowledge of its habits and manners are so imperfect, that of its nidification and eggs we can give no certain information.

The sexes offer but little difference of colouring: the young, on the contrary, before the second or third year, at which period they attain their adult colouring, are much more obscure in their tints, and exhibit none of that metallic lustre which afterwards forms so characteristic a feature.

The adult birds have the head of a dark chestnut; the neck, breast, top of the back, the upper edge of the wing, and all the under parts of a rich reddish chestnut; the lower part of the back, the rump, quill- and tail-feathers of a dark green, with bronze and purple reflections; the naked skin round the eyes olive green, becoming more grey towards the outer margin; the irides brown; legs and feet dull olive brown.

In the young, the feathers, which are of a fine chestnut in the adults, exhibit faint indications only of this colour, being of a dull brown, and each feather on the neck is edged with a margin of greyish white; the other parts display but little of the metallic reflections.

The Plate represents an adult male about three fourths of the natural size.

COMMON CURLEW.
Numenius arquata. *(Lath.)*

Genus NUMENIUS, *Lath.*

Gen. Char. *Bill* long, slender, incurved, slightly compressed, rounded through its whole length, with the tip of the upper mandible projecting beyond the lower one; hard, and semi-obtuse; laterally furrowed for three fourths of its length. *Nostrils* basal, placed in the lateral groove, linear, and covered above by a naked membrane. *Lores*, or space between the bill and eyes, covered with feathers. *Legs* long, slender, naked above the tarsal joint. *Feet* four-toed; three before and one behind; the front ones connected at the base by a large membrane. *Toes* short; the outer and inner ones of nearly equal length; hind toe short, and articulated above the plane of the others, upon the tarsus, its tip only resting upon the ground. *Claws* short and blunt. Front of the tarsi partly scutellated; back reticulated. Front of the toes scutellated.

COMMON CURLEW.

Numenius arquata, *Lath.*

Le Grand Courlis cendré.

Since the North American Curlew (*Numenius longirostris*, Wils.) has been found to possess good specific differences, the range of the present species will be restricted to the regions of the Old World; and different as the climates of this vast range must necessarily be, the Common Curlew is found equally diffused from the sultry portion of the torrid zone to the frozen countries of the North: the islands of the Pacific Ocean, particularly New Holland, are not devoid of its presence, and we also possess examples from China, Nepaul, &c. In its disposition the Common Curlew is extremely wary and distrustful; and it resorts to such wild and open situations that the greatest ingenuity is required to approach it, which, indeed, is seldom accomplished except by the sportsman secreting himself in the neighbourhood of its haunts, and thereby obtaining an opportunity of shooting it while flying over the place of his concealment.

It is migratory in its habits; at least those which frequent the temperate portions of Europe pass the winter on the sea-coast and the neighbouring marshes, and retire to the high lands of Norway and Sweden during the months of summer, such situations being conducive to its security during the period of incubation. From the circumstance of a few of these birds being left on most of our extensive moors and wild open districts during the breeding-season it may be considered a permanent resident in England, although the greater number of those which winter on our shores do certainly retire northwards to the Western, Orkney, and Shetland Islands, whence probably many of them cross the Channel to Norway and Sweden. The Common Curlew possesses extraordinary powers of flight, and is consequently enabled easily to pass from the shores of the sea, at every rising tide, to inland wilds, fields, morasses, &c., and by some peculiar instinct to return again to the coast almost at the moment of the commencement of the ebb, when it follows the receding waves, and feeds upon such marine worms, crustacea, &c. as are left on the sands. We cannot refrain from here inserting an interesting note on the habits of this bird by Sir Wm. Jardine, Bart., copied from his edition of Wilson's American Ornithology.

The Common Curlew, "*Numenius arquata*, during the breeding-season, is entirely an inhabitant of the upland moors and sheep-pastures, and in the soft and dewy mornings of May and June forms an object in their early solitude which adds to their wildness. At first dawn, when nothing can be seen but rounded hills of rich and green pasture, rising one beyond another, with perhaps an extensive meadow between, looking more boundless by the shadows and mists of morn, a long string of sheep marching off at a sleepy pace on their well-beaten track to some favourite feeding-ground, the shrill tremulous call of the Curlew to his mate has something in it wild and melancholy, yet always pleasing to the associations. In such situations do they build, making almost no nest, and, during the commencement of their amours, run skulkingly among the long grass and rushes, the male rising and sailing round, or descending with the wings closed above his back, and uttering his peculiar quavering whistle. The approach of an intruder requires more demonstration of his powers, and he approaches near, buffeting and *whauping* with all his might. When the young are hatched, they remain near the spot, and are for a long time difficult to raise; a pointer will stand and road them, and at this time they are tender and well-flavoured. By autumn they are nearly all dispersed to the sea-coasts, and have now lost their clear whistle."

The sexes are alike in plumage, and their flesh is by many considered as a great delicacy for the table.

Bill blackish brown for half its length from the tip, the rest being fleshy white; head, neck, and upper surface light grey, the centre of each feather being dark brown; rump white; tail transversely barred with brown and white; quills dark brown, the shafts being white, and the inner webs barred with the same colour; throat, neck, and chest yellowish white thickly striped with olive brown; rest of the under surface white; legs and feet bluish lead colour; irides dark brown.

The Plate represents an adult male of the natural size.

WHIMBREL.
Numenius Phæopus, (Lath.)

Drawn from Nature & on stone by J & E Gould.

Printed by C Hullmandel.

WHIMBREL.

Numenius Phæopus, *Lath.*

Le Courlis corlieu.

ALTHOUGH North America presents us with a species closely allied to the Whimbrel, and with which it has been confounded by some writers, we believe we are safe in affirming that the British species is limited in the range of its habitat exclusively to the Old World, over which it appears to be very widely dispersed : we have received it in abundance from the Himalaya, and several other parts of India, as well as from Northern Africa. In the temperate latitudes of Europe, and doubtless in those of Asia, it is strictly a winter visitant, retiring on the approach of spring to the regions within the arctic circle, where it incubates and passes the summer, and where the land, almost destitute of inhabitants and abounding in extensive flats and morasses, appears to be peculiarly suited for the summer sojourn of vast numbers of the *Scolopacidæ* and other Grallatorial birds.

The inferiority of its size will always distinguish the present species from the Curlew, which in other respects it closely resembles.

In the British Islands it is tolerably common throughout the winter, inhabiting all the low flat parts of our coasts, and especially the mouths of our larger rivers, and feeding upon various molluscous and other marine animals, which it takes when the tide is at the ebb, and retires to the neighbouring saline marshes when the water covers the shore. It is generally seen in small flocks, which on being approached take wing and fly off with great vigour and rapidity. Dr. Fleming states that it has been known to breed in Shetland, constructing its nest on the exposed heath and moorlands. Its eggs, four in number, are of an olive brown colour, blotched and spotted with darker reddish brown ; but as far as our own researches go we have never been able to meet with an example of its eggs, or an instance of the young being obtained.

The sexes are alike in plumage, and differ but little in size or in the winter or summer livery.

The bill is black with the base of the under mandible flesh colour ; top of the head brown with a longitudinal stripe of greyish white down the centre ; throat, rump, and abdomen white ; cheeks, stripe over the eye, and chest greyish white with a longitudinal dash of dusky brown on each feather ; back and wings mottled with deep brown and greyish white ; tail thickly barred with brown and white ; legs brownish black.

We have figured an adult of the natural size.

SLENDER-BILLED CURLEW.
Numenius tenuirostris. (*Sav.*)

Drawn from Nature & on Stone by J.& E.Gould.

Printed by C.Hullmandel.

SLENDER-BILLED CURLEW.

Numenius tenuirostris, *Savi.*

INDEPENDENTLY of the size of the bill, the distinct spotting of the breast will at all times serve to distinguish this bird from its near allies the Curlew and Whimbrel, which until lately were the only examples of the genus *Numenius* found in Europe: the honour of adding the present very elegant species is due to Professor Savi. It is a native of the southern portions of Europe, and, which is very singular, we have never seen it from Asia or Africa, countries in which both the common species, the Whimbrel and Curlew, are very abundant.

Of its habits, manners, &c. nothing is known, but they doubtless resemble those of the other members of the genus.

Throat white; the whole of the head and the remainder of the neck pale brown, each feather ornamented in the centre near the tip with an oblong mark of deep brown; upper part of the back and lesser wing-coverts deep brown margined with pale brown; greater wing-coverts and the long scapularies pale brown with whitish edges, and barred with deep brown; tail white, barred with blackish brown; breast and abdomen white passing into buff on the flanks, and each feather having a large heart-shaped mark in the centre, near the tip; vent and under tail-coverts white; base of the lower mandible reddish, the remainder and the whole of the upper mandible black; legs and feet greenish black.

We have figured an adult of the natural size.

BLACK-TAILED GODWIT.
Limosa melanura, /teacher/.

Genus LIMOSA.

GEN. CHAR. *Bill* very long, rather thick at the base, compressed, more or less turned upwards; both mandibles laterally grooved to within a short distance of the point, which is
somewhat dilated and blunt; tip of the upper mandible projecting beyond the lower one.
Nostrils basal, placed in the lateral groove, narrow and longitudinal. *Wings* acuminate,
of mean length, the first quill-feather the longest. *Legs* long and slender; a great part of
the tibiæ naked; front and back part of the tarsi scutellated. *Feet* four-toed, three before
and one behind; the outer toe united to the middle one by a membrane as far as the first
joint; the inner one nearly free; hind toe short, and articulated upon the tarsus.

BLACK-TAILED GODWIT.

Limosa melanura, *Leisl.*

Le Barge à queue noire.

THIS stately bird, one of the finest of its race, was so common in the low lands and fenny districts of England
so late as twenty years since as to have been then regarded as one of the very commonest of the *Scolopacidæ*
visiting the British Isles; since which period, however, its numbers appear to have gradually diminished, so
much so that a fresh-shot native specimen is now considered an article of a somewhat singular nature.

The Godwits are subject to considerable changes in their plumage, but in no one species are they more
striking than in the Black-tailed, its winter dress consisting of grey above and white beneath, while the spring
or nuptial attire is altogether different, the bird being then characterized by a much brighter and more gay
colouring. From the periodical changes to which both sexes are subject, much confusion has arisen in the
works of the older ornithologists, which by more recent investigation has been entirely cleared up, and the
fact is now completely established, that the birds exhibiting variations of plumage between those here
figured are identically the same; and as the birds of this genus do not acquire the rich colouring of the
summer plumage until they are two or three years old, it is not surprising that the succession of intermediate
stages which they continually present should have misled the ornithologist. The most clear and certain mark
by which the present species may at all times be distinguished from its near ally, is the colouring of the
tail, which is perfectly black for half its length from the tip, while, as its name implies, the barred tail of
the other species is a feature equally distinguishing.

Besides the European continent, over the whole of which it is distributed, we find examples of the Black-
tailed Godwit in most of the collections from Africa and India. It is especially abundant in Holland, whence
it is brought to the London market in great abundance, both living and dead. In its manners and actions it
is elegant and graceful; and soon becoming tame and familiar, it forms an agreeable tenant for the aviary.
The flesh of the adult is rather coarse and rancid, but the young of the year are more delicate, and are
therefore more in request for the table. A few pairs annually resort to the marshes in the neighbourhood
of Yarmouth, and to the fens of Lincolnshire, but they are rarely permitted to breed unmolested, their large
size and peculiar actions being sure to attract the notice both of the sportsman and the egg-gatherer. The
eggs are four in number, of an olive green faintly blotched with black, and are deposited on the bare ground
among the herbage, with little or no nest.

Its food consists of worms, shell-snails, insects, and their larvæ, for procuring which its lengthened tarsi
and greatly developed bill admirably adapt it. It runs with great facility, and its power of flight is such as
the lengthened form of its wing would lead us to expect.

The female surpasses the male in size, and frequently in the brilliant colouring of the summer plumage.

In winter the whole of the upper surface is brownish grey; the tail for half its length from the tip black;
under surface greyish white; primaries blackish brown; legs and feet black tinged with olive.

In summer the crown of the head, the throat, neck, and breast are clear reddish brown streaked and
barred with black; the whole of the upper surface and flanks transversely barred and edged with black and
red; base of the bill red, deepening into black at the tip.

The young of the year has the crown of the head blackish brown, each feather being edged with reddish
brown; the front of the neck and chest clear reddish ash; a streak between the bill and the eye, and the
whole of the under-surface white; wing-coverts grey, margined and terminated with reddish white.

The Plate represents two birds, one in the winter, the other in the summer plumage, of the natural size.

BAR-TAILED GODWIT.
Limosa rufa. (Bris.)

BAR-TAILED GODWIT.

Limosa rufa, *Briss.*

La Barge rousse.

NOTWITHSTANDING multitudes of this bird visit our island during their vernal and autumnal migrations, we possess no authenticated instance of its having remained and bred with us, nor even, indeed, is the colour of its eggs satisfactorily ascertained. Like many others of the same tribe, they appear to make our island a resting-place during their periodical flights only: we should observe, however, that a limited number sojourn with us during the winter, the number being increased in spring by large flocks which have wandered further south-wards; the whole then return to their northern home. That Iceland, Lapland, and the regions within the arctic circle are the summer abode of the Bar-tailed Godwit, is a fact not to be questioned; and from thence they retire in autumn divested of that richly coloured livery with which they are adorned at their departure from the South in spring. Large flocks of the young also pass over at the time of the arrival of the adults, or soon after: these are the progeny of the past summer, and are to be distinguished by a more spotted and streaky plumage, and by their having the feathers of the whole of the upper surface margined with a lighter colour; the breast is then of a dull fawn yellow strongly streaked with brown.

With regard to the adult birds, few species present a more decided contrast in the colours of their summer and winter plumage; from which circumstance has arisen much confusion and a perplexing list of synonyms in the works of various writers; indeed, both the males and the females, as well as the young, have been charac-terized as distinct species, nor was it until very lately that these errors were corrected.

The places most frequented by the Bar-tailed Godwit during its residence in our island are the low muddy shores of the sea, the borders of creeks and inlets, saline marshes along the coast, and the mouths of large rivers: here, with other Waders, they may be observed in small flocks busily searching for their food, which consists of aquatic insects, worms, and mollusca. They run with great facility over the oozy ground, and fly, when roused, to a considerable distance, uttering as they rise on the wing a hoarse deep note. Their visits to the Continental districts take place at the same time as in our island. In Holland and the level parts of France, which afford them a congenial residence, they abound much more than in England, and, like the Black-tailed Godwit, are annually killed in great numbers and forwarded to the London markets.

Both sexes assume the red plumage in spring, but it is much deeper and more conspicuous in the male: the female is, however, considerably larger in size, as we have seen is the case with its near ally and many others of its family.

In summer the crown of the head and back of the neck are rufous, streaked longitudinally with blackish brown; the whole of the upper surface blackish brown mottled with rufous; quills dark brown; throat and under surface deep rufous; rump white; tail barred with reddish white and black; feet and legs dark olive; bill reddish yellow at the base and dark brown at the point.

All the parts which are red in summer are greyish white in winter, and the dark markings of the upper surface are exchanged for more obscure tints of brown.

The Plate represents two birds, one in the summer and the other in the winter plumage, of the natural size.

TEREK GODWIT.
Limosa Terek. (Temm.)

Drawn from Nature & on Stone by J & E Gould.

Printed by C Hullmandel.

TEREK GODWIT.

Limosa Terek, *Temm.*

La Barge Terek.

In figuring this bird as an occasional visitant of the continent of Europe we are entirely guided by our friend M. Temminck, who has not only forwarded us a specimen to figure from, but in a letter accompanying it states that an example has been killed in Normandy ; and on comparing the European-killed specimen with others from Borneo and Japan, no difference exists between them. The specimen forwarded to us is in its winter plumage, which is here represented, and which gives place to a mottled and spotted plumage during the spring and summer, particularly on the upper surface, where the markings are larger and assume a lanceolate form. Although we have followed M. Temminck in placing this elegant species with the God-wits, we are not fully persuaded that its situation is natural. We ourselves, notwithstanding the upward curvature of the bill, are inclined to believe it to be nearly allied to the true *Tringas*, or Sandpipers ; but as an acquaintance with its habits, manners, general economy, and mode of flight would alone enable us to confirm our opinion respecting its natural situation, we leave it where it has been placed by M. Temminck.

The head, back and sides of the neck, all the upper surface, and tail pale brown with a fine line of a darker tint down the centre of each feather ; shoulders and primaries dark brown, with the shaft of the first quill white ; secondaries, chin, front of the neck, and all the under surface pure white ; sides of the upper mandible and base of the lower yellow ; the remainder of the bill dark brown ; legs and feet yellow.

The figure is of the natural size.

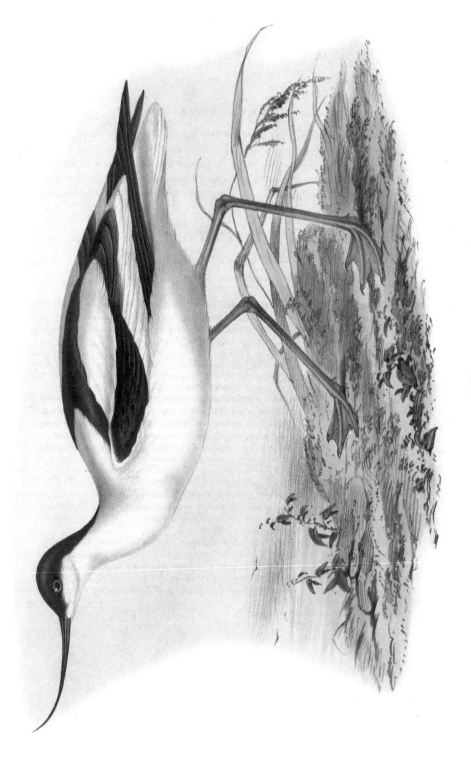

AVOCET.
Recurvirostra avocetta. (Linn.)

Printed by C.Hullmandel

Genus **RECURVIROSTRA,** *Linn.*

GEN. CHAR. *Beak* long, slender, feeble, depressed through its whole length; the point flexible, and turning upwards; the *upper mandible* grooved along its surface, the *under* grooved laterally. *Nostrils* on the surface of the beak, long and linear. *Tarsi* long and slender. *Toes* three before, palmated as far as the second articulation; and one behind, merely rudimentary and articulated high on the tarsus. *Wings* pointed; first *quill-feather* longest.

AVOCET.

Recurvirostra avocetta, *Linn.*

L'Avocette à nuque noire.

THE very interesting and well-defined genus *Recurvirostra* contains but a very limited number of species, of which the present is the only European example; not, however, that it is altogether confined to that portion of the globe, but is also found from Egypt throughout the whole of Africa, to its most southern boundary, as is proved by the identity of individuals killed on the northern coast, and at the Cape of Good Hope. It is found in India also, although rather sparingly. It would appear, however, that Holland, France and Germany may be considered its natural and most congenial habitat; preferring in each of these countries the low flat lands bordering the sea, salt marshes and swamps occasionally covered by the tide. A century ago, before our fens were drained, and while extensive marshes afforded food and concealment, the Avocet was common in England, frequenting in abundance the fens of Lincolnshire and Norfolk, to which it even now occasionally resorts for the purpose of incubation. We believe it to be strictly migratory, arriving in our latitudes only at those seasons when the marshes and lakes are unfrozen and abound in its peculiar food, which consists chiefly of minute insects, the larvæ of crustacea, &c., for the taking of which its beak is most singularly and beautifully adapted; nor is the construction of its legs less adapted to sustain it on the mud and swampy ground in which it wades; its semipalmated feet being more adapted for the purpose of supporting its weight on a soft and yielding surface, than for assisting it when swimming, to which it seldom resorts but in cases of necessity. The places it selects for the purpose of incubation, are similar to those of other marsh birds, usually a depression in the ground, making little or no nest, where it deposits its eggs, which rarely exceed two in number, and which, except in size, so nearly resemble in shape and markings those of the Lapwing as to be easily mistaken for them. They are however much larger, measuring 2 inches 1 line in length by 1 inch 7 lines in width, of an olive brown, spotted with black.

Dr. Latham informs us that the Avocet is very bold in defence of its young, and when disturbed in the breeding season, it hovers over the sportsman's head like the Lapwing, and flies with its legs and neck extended, uttering a sharp note like the word *twit twit* often repeated. The young soon resemble the adults in colouring, and old birds present no external differences, having the whole of the body white, with the exception of the top of the head, the back of the neck, the scapulars and quill-feathers, which are black; beak black; irides reddish brown; feet and legs blueish ash.

We have figured an adult bird in full plumage.

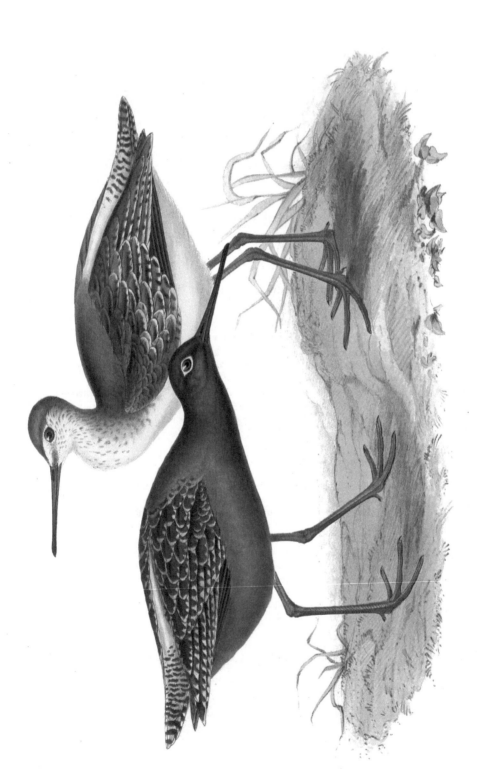

SPOTTED REDSHANK.
Totanus fuscus. *(Leisl.)*

Genus TOTANUS.

GEN. CHAR. *Beak* moderate, often slightly recurved, rounded, hard, and pointed; the upper mandible sulcated, and having the tip arched and curving over that of the lower one. *Nostrils* basal, lateral, linear, and longitudinally cleft in the furrow of the mandible. *Legs* long, slender, naked above the tarsal joint. *Toes* three before and one behind, the former united at the base by a small membrane, hind toe short; front of the tarsus and toes scutellated.

SPOTTED REDSHANK.

Totanus fuscus, *Leisl.*

Le Chevalier arlequin.

THE ornithologist cannot but be gratified and greatly interested in the study of the family of which the present species forms an example. Independently of the graceful form which this elegant bird possesses, it is characterized by a change of colouring as singular as it is chaste and becoming. We allude to the contrast of its summer and winter dress, as well as to that which distinguishes the young of the year, in which stage the whole of the upper plumage has the margins of the feathers thickly spotted with white, as is found in the wing-feathers of the birds in the present Plate, more particularly like that of the olive-coloured bird, which is figured in a particular stage, partly that of winter and partly that of immaturity, the uniform colouring of the back, which in perfectly adult birds extends over the wings, being the usual characteristic of the winter dress. The dark-coloured bird in the foreground of our Plate is in a stage of plumage common to both sexes during the breeding-season, in which state they are extremely rare, their breeding-places being in the northern regions of Europe, and the examples we occasionally obtain are consequently in general more or less imperfect: notwithstanding this, we have had frequent opportunities of examining this bird in all its changes; it is, however, one of those irregular visitants whose appearance is not to be depended on. It traverses an extensive range of country, being abundant in many parts of Asia, whence we have received examples strictly identical with those of Europe. Its favourite places of residence are the borders of rivers, lakes, and morasses, where it feeds on freshwater mollusca, insects, and worms. On the neighbouring parts of the Continent it appears to be as scarce and irregular in its visits as it is in the British Islands. Of its nest and eggs nothing has been correctly ascertained.

The plumage of summer may be thus detailed : Head, neck, back, and under parts dark greyish black; the rump white; wing-coverts and scapularies dark greyish black, with their edges spotted with numerous dots of white; upper tail-coverts barred with black and greyish white; quills black; bill black at the tip and red at the base; legs orange red.

In winter the whole of the upper surface is brown, with a tinge of olive; the under surface is pure white; the legs, beak, wings, and tail being the same as in the summer plumage.

The sexes offer no difference except in size, the female being somewhat the largest.

The Plate represents two birds, the one in the plumage of summer, the other a bird between youth and maturity, assuming its winter dress.

REDSHANK.
Totanus calidris. (*Bechst.*)

Drawn from Life & on Stone by H.E.Smith.

Printed by C.Hullmandel.

REDSHANK.

Totanus calidris, *Bechst.*

Le Chevalier Gambette.

THIS species of Redshank is not so remarkable for the transitory changes in the colour of its plumage, as is its allied congener the *Totanus fuscus*; for while this last would appear to form another species at different seasons of the year, if change of colour was sufficient, the present bird has the plumage very generally spotted during the seasons of spring and summer, the ground-colour only remaining wholly unchanged. In point of number the Redshank is by far the most common, and is very universally spread over the marshy and low lands of Europe. It is indigenous to the British Isles, and is equally dispersed from Orkney and Shetland to our most southern counties. During the autumn and winter its favourite localities are the edges of the sea and mouths of large rivers, running with great ease and elegance over the flat muddy plains which have been recently left bare by the retiring tide. In the summer it takes to the adjacent marshes, where amid tufts of grass or rushes it constructs a slight inartificial nest, in which it deposits four eggs, rather larger than those of the Snipe, of greenish yellow marked with brown spots which blend together at the larger end. Although the young are soon able to run and provide for themselves, they are not in possession of the power of flight for a considerable period; when disturbed they hide themselves among the herbage with the utmost caution, while the parents may be observed at a distance uttering their querulous and pitiful notes, not unfrequently perched on some neighbouring post or rail, where, with drooping wings and outspread tail, they display the most grotesque and singular appearance.

The sexes offer no difference of plumage, but if compared together may be distinguished by the larger size of the female.

The colour during the spring and breeding season is as follows.—From the eyes to the beak an obscure white mark; the head, back of the neck, top of the back, scapulars and wing-coverts of a greyish olive-brown; on each feather there is a large longitudinal brown mark, except on those of the scapulars and wing-coverts, where there are small black transverse bars; the rump white; the sides of the head, the throat, and all the underparts white, each feather having a longitudinal dash of brownish black which becomes oblique on the abdomen and under tail-coverts; the feathers of the tail are barred with black and white terminating in the latter, the white portion of the four middle ones being tinged with ash colour; the basal half of the beak and the feet are of a bright orange red. As winter comes on these markings become more and more obscure, till at length the back of the neck and the whole of the upper surface are of one uniform ashy brown; the throat, the sides of the head, the fore-part of the neck, and breast, of a greyish white, each feather having the shaft of a dull brown; the rump and underparts of a pure white; the tarsi of a pale reddish orange; irides brown.

Length ten inches.

We have figured adults in the summer and winter plumage.

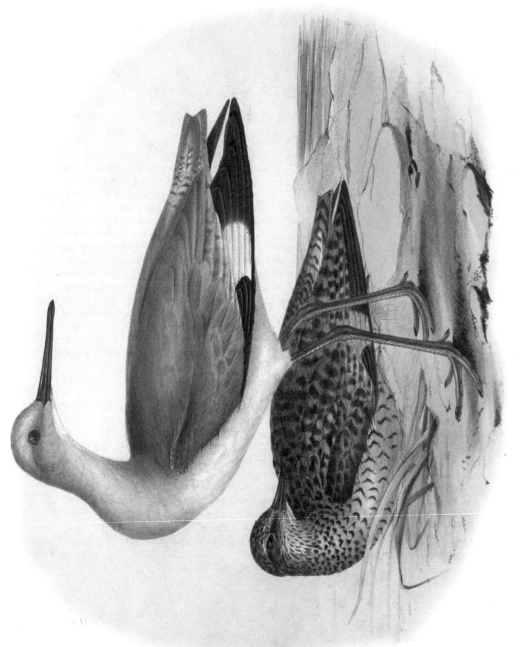

SEMIPALMATED SANDPIPER.
Totanus semipalmatus. /Temm./

SEMIPALMATED SANDPIPER.

Totanus semipalmatus, *Temm.*

Le Chevalier semi-palmé.

THE celebrated Wilson having beautifully portrayed the history of this species, we have taken the liberty of extracting rather largely from his valuable work, before which, however, we would state that it is on the authority of the continental naturalists that we have been induced to give a figure of it in the present work. M. Temminck informs us that it is accidentally found in the North of Europe, but like ourselves, quotes from Wilson an account of its food, manners, &c. We have also been favoured with an European-killed specimen presented to us by Professor Lichtenstein of Berlin; no doubt therefore can exist as to the propriety of admitting it into the Fauna of this portion of the globe, although America must be considered as its true habitat.

" This," says Wilson, " is one of the most noisy and noted birds that inhabit our salt marshes in summer. Its common name is the Willet, by which appellation it is universally known along the shores of New York, New Jersey, Delaware, and Maryland, in all of which places it breeds in great numbers. It arrives from the south on the shores of the middle states about the 20th of April or beginning of May; and from that time to the last of July its loud and shrill reiterations of *pill-will-willet, pill-will-willet*, resound almost incessantly along the marshes, and may be distinctly heard at the distance of more than half a mile. Their nests are built on the ground, among the grass of the salt marshes, and are composed of wet rushes and coarse grass, forming a slight hollow or cavity in a tussock. This nest is gradually increased during the period of laying and sitting to the height of five or six inches. The eggs are usually four in number, very thick at the great end, tapering to a narrow point at the other, and of a dark dingy olive, largely blotched with blackish brown, particularly at the great end. The eggs, in every instance that has come under my observation, are placed during incubation in an almost upright position, with the large end uppermost; and this appears to be the constant practice of several other species of birds that breed in these marshes. During the laying season, the Crows are seen roaming over the marshes in search of eggs, and wherever they come spread consternation and alarm among the Willets, who in united numbers attack and pursue them with loud clamours."

The Willet subsists chiefly on small shell-fish, marine worms, and aquatic insects, in search of which it regularly resorts to the muddy shores and flats at low water.

This species differs considerably in its summer and winter plumage, the latter being of a pale dun colour with darker shafts, and the former as follows:

Upper surface dark olive brown, each feather streaked down the centre and crossed with irregular lines of black, and numerously blotched with dull yellowish white; wing-coverts greyish olive; basal half of the primaries white, the remainder black; secondaries white; rump dark brown; upper tail-coverts white, barred with olive; tail pale olive crossed with bars of dark brown; chin white; breast and flanks cream colour transversely mottled with olive; belly and vent white, the latter barred with olive; tip of the bill black; the base and the legs and feet pale lead colour.

We have figured two birds of the natural size, one in the summer and the other in the winter plumage.

GREEN SHANK.
Totanus glottis. (Bechst.)

Drawn from Nature & on Stone by J & E Gould.

Printed by C Hullmandel

GREENSHANK.

Totanus glottis, *Bechst.*

Le Chevalier aboyeur.

WE are not inclined to consider the upward curvature which the mandibles of this bird exhibit of sufficient importance to warrant its separation from the genus *Totanus*, answering as the rest of its characters do to those upon which that genus was established; for, like most of the species, it undergoes a slight periodical change in the colouring of its plumage, the summer livery, which is varied with markings of dark grey, particularly on the chest and flanks, giving place in autumn and winter to a uniform tint of white over the whole of the under surface. It is in this latter stage that the bird is represented on our Plate. In its habits and manners also, as well as in the circumstance of the sexes not being distinguished by the colouring of their plumage, it is strongly allied to the two species of Redshank, which may be considered as typical examples.

From the circumstance of the Greenshank having been lately added to the Fauna of America by that justly celebrated ornithologist M. Audubon, it may be considered to possess a range scarcely equalled in extent by any of the Sandpipers, as it is generally spread over the whole of India and Africa, nor is it less numerously diffused over the countries of Europe. Although it does not make the British Islands either a place of permanent residence or of incubation, still it is sufficiently numerous during its vernal and autumnal migrations to be considered as strictly within the list of British species. In England it frequents the shores of the sea and the mouths of the larger rivers: on the Continent, during the winter, it is common on the coast of Holland, but less so on that of France; it also occurs on the lakes of Switzerland and Germany, as well as on the banks of the Rhine and other large rivers.

Its breeding-place must in all probability be looked for in the high northern latitudes, which form a place of summer residence to so many other members of the family.

Its food consists of aquatic worms, insects, mollusca, and the small fry of fishes.

Top of the head and sides of the neck dull white, streaked with dusky brown; face, sides of the head, throat, front of the neck, and all the under surface pure white; flanks streaked and rayed transversely with light brown; upper part of the back, wing-coverts, tertials, and scapulars deep brown, tinged with purple; each feather margined with greyish white, and dotted with brown of a deeper tint; quills brownish black; the shafts of the first being white; lower part of the back and rump white; tail white irregularly barred with brown; bill brownish black; legs and feet greenish grey.

The figure is of the natural size.

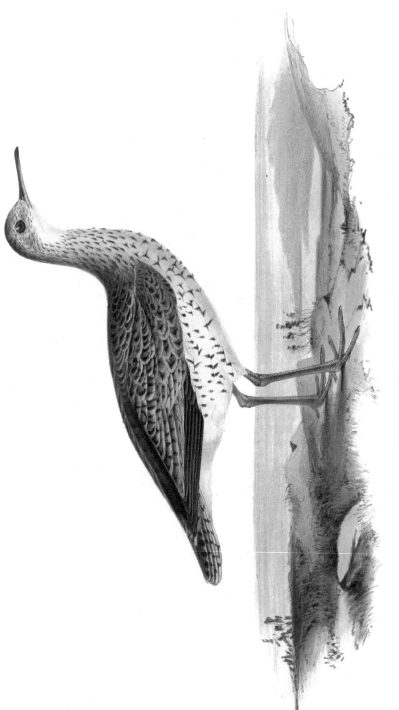

BARTRAM'S SANDPIPER.
Totanus Bartramia. *(Temm.)*

Drawn from Nature & on Stone by J & E Gould.

Printed by C. Hullmandel.

BARTRAM'S SANDPIPER.

Totanus Bartramia, *Temm.*

Le Chevalier à longue queue.

For the history of this bird we must have recourse to the valuable pages of the justly celebrated American ornithologist Wilson, by whom it was first discovered and described. Its occurrence in Europe is so rare that we know of only two or three instances of its having been procured within the limits of this portion of the globe. M. Temminck informs us that one was killed in autumn on the Dutch coast, and that another came under his observation which had been taken in the eastern part of Germany. We believe that as yet it has never been seen in Great Britain.

"Unlike most of their tribe," says Wilson, "these birds appear to prefer running about among the grass, feeding on beetles and other winged insects. There were three or four in company; they seemed extremely watchful, silent, and shy, so that it was always with extreme difficulty I could approach them.

"Having never met with them on the sea shore, I am persuaded that their principal residence is in the interior, in meadows and suchlike places. They run with great rapidity, sometimes spreading their tail and dropping their wings, as birds do who wish to decoy you from their nest: when they alight they remain fixed, stand very erect, and utter two or three sharp whistling notes as they mount to fly. They are remarkably plump birds, weighing upwards of three quarters of a pound; their flesh is superior, in point of delicacy, tenderness, and flavour, to any other of the tribe with which I am acquainted.

"This species is twelve inches long and twenty-one in extent; the bill is an inch and a half long, slightly bent downwards, and wrinkled at the base; the upper mandible black on the ridge, the lower as well as the edge of the upper of a fine yellow; front, stripe over the eye, neck, and breast pale ferruginous marked with small streaks of black, which on the lower part of the breast assume the form of arrow-heads; crown black; the plumage slightly skirted with whitish; chin, orbit of the eye, whole belly and vent pure white; hind head and neck above ferruginous, minutely streaked with black; back and scapulars black, the former slightly skirted with ferruginous, the latter with white; tertials black bordered with white; primaries plain black; shaft of the exterior quill snowy, its inner vane elegantly pectinated with white; secondaries pale brown spotted on their outer vanes with black and tipped with white; greater coverts dusky, edged with pale ferruginous and spotted with black; lesser coverts pale ferruginous, each feather broadly bordered with white, within which is a concentric semicircle of black; rump and tail-coverts deep brown black, slightly bordered with white; tail tapering, of a pale brown orange colour beautifully spotted with black, the middle feathers centred with dusky; legs yellow tinged with green; the outer toe joined to the middle by a membrane; lining of the wings elegantly barred with black and white; iris of the eye dark or blue black; eye very large. The male and female are nearly alike."

We have figured an adult male of the natural size.

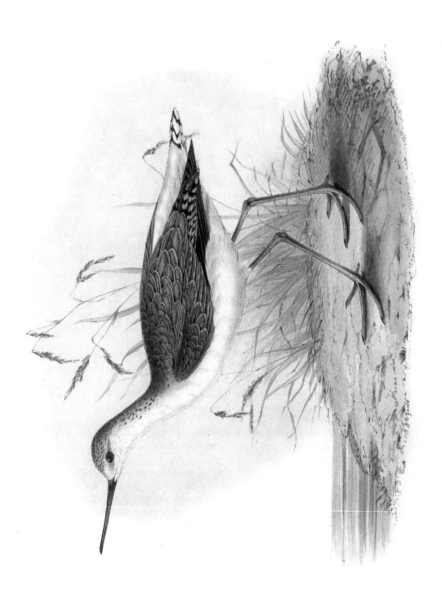

MARSH SANDPIPER.

Totanus stagnatilis. (Bechst)

Drawn from Nature & on Stone by J&E Gould.

Printed by C. Hullmandel.

MARSH SANDPIPER.

Totanus stagnatilis, *Bechst.*

Le Chevalier stagnatile.

THIS elegant species of Sandpiper is here represented in its adult state, and in the plumage of winter, which differs from that of summer only in being more generally light on the upper surface, the ash-grey being uninterrupted with transverse bars of black, the last remains of which may be observed on the scapularies : the under surface is also more purely white, with scarcely any traces of the longitudinal black streaks which characterize the plumage of summer. In form and general contour, the Marsh Sandpiper approaches closely to the Spotted Redshank (*Totanus fuscus*, Leisl.), but is not much more than half its size ; besides which, the legs of the former are of an olive-green, while those of the latter are red.

The Marsh Sandpiper is not a native of the British Islands, and it appears to be extremely rare upon the Continent : its extent of habitat, however, is by no means very limited, since, according to M. Temminck, who favoured me with the examples from which the present figure was drawn, it migrates from the North of Europe, its summer habitat, throughout the eastern provinces as far as the Mediterranean, frequenting the borders of large rivers, but never taking up its residence on the sea-shore.

As its name implies, it obtains its food from the marshes and humid tracts which border inland rivers and sheets of fresh water.

The top of the head grey, with longitudinal spots of black ; upper surface grey, each feather having a lighter margin ; wings blackish brown ; tail white, marked with diagonal bars of brown ; the other parts and the whole of the under surface white ; the bill and irides blackish brown ; the legs bright olive-green : such is the plumage of the adult in winter. Its total length is about nine inches.

The young have the whole of the upper plumage of a brownish black, each feather bearing a yellowish border ; the scapularies barred with narrow diagonal lines of black ; and the tarsi of a greenish ash colour.

The figure is of the natural size.

WOOD SANDPIPER.
Totanus Glareola. *(Temm.)*

GREEN SANDPIPER.
Totanus ochropus. *(Temm.)*

Drawn from Nature & on Stone by J. & E. Gould.

Printed by C. Hullmandel.

GREEN SANDPIPER.

Totanus ochropus, *Temm.*

Le Chevalier Cul blanc.

ALTHOUGH we believe that this delicate species frequently passes the summer in Wales and other mountainous districts of Great Britain, and consequently breeds there, we are not in possession of sufficient facts to warrant us in stating positively that such is the case. We are strengthened in this opinion, however, by the circumstance of Mr. Doubleday, an accurate observer of nature, having informed us that he has observed it flying over the smaller streams in the neighbourhood of Snowdon in the middle of summer. It generally arrives in Britain at the latter end of September, when it disperses either singly or in pairs over nearly every part of our island, more particularly its interior portions, but always in very small numbers. It is most frequently met with during its autumnal and spring migrations, and appears to give preference to the edges of small ponds, rivulets, and ditches, particularly those which are secluded: its habits in this respect differ very considerably from those of most of the other members of its family, which resort to the sea-beach and the sides and mouths of large rivers as favourite places of abode.

The snow-white rump of the Green Sandpiper renders it a conspicuous object when flushed, at which time it utters a shrill whistling note, from which circumstance it is known to many persons, particularly sportsmen, by the name of the Whistling Snipe: it runs with great activity, but generally flies low, skimming over the surface of the water, and following the bends and angles of the stream. It possesses an extensive range of habitat, being dispersed over the whole of Europe, the greater part of Asia, and Africa; but it is not found in America, as most naturalists have asserted it to be, its place there being supplied by a nearly allied but totally distinct species. It differs from the Wood Sandpiper in its larger size, its shorter tarsi, and in the more diminutive spotting of the upper surface.

According to M. Temminck, the nest is made by the side of a stream, and the eggs are greenish white blotched with brown.

The sexes are so nearly alike in size and colour that a description of both is quite unnecessary.

The top of the head, back of the neck, and upper surface olive brown; the back, the scapularies, and most of the wing-feathers marked at their edges with minute spots of yellowish white; quills dark brown; sides of the chest and flanks longitudinally streaked and spotted with brown; rump, under surface, and tail white, the latter having the four central feathers strongly barred with black; feet and legs olive; bill black with a tinge of olive; irides dark brown.

WOOD SANDPIPER.

Totanus Glareola, *Temm.*

Le Chevalier sylvain.

THIS species has been so frequently confounded with the preceding, that we are induced to figure both on the same Plate, in order to enable our readers more readily to distinguish their differential characters. There is no difference in the colouring of the sexes of either species, and as their plumage is not influenced by the seasons or other causes, we trust our Plate will illustrate every feature necessary to render their distinctness sufficiently apparent.

The Wood Sandpiper is still more rarely seen in the British Islands than its near ally, but in every other particular the history of the preceding species is applicable to the present; it is, however, even more widely dispersed, as is proved by its being found not only over the whole of the Asiatic continent, but in most of the islands of the Pacific Ocean also, which we believe is not the case with the Green Sandpiper.

A streak of brown passes between the bill and the eyes; crown of the head and sides of the neck brown, streaked with a darker tint; whole of the breast and scapularies dark brown, the edge of each feather being spotted with greyish white; lower part of the rump and the tail white, the latter numerously barred with brown; legs and feet dark olive; bill black; irides dark brown.

We have figured each species of the natural size.

COMMON SANDPIPER.
Totanus Hypoleucos. (Temm.)

Drawn from Nature & on Stone by J.& E. Gould.

Printed by C. Hullmandel.

Genus TOTANUS, *Bechst.*

GEN. CHAR. *Bill* long, rounded, solid, hard, and drawn to a point; in some species slightly incurved; upper mandible sulcated, the furrow generally about half the length of the bill; the tip arched and curving over that of the lower one; tomia of both mandibles bending inwards progressively to the point. *Nostrils* basal, lateral, linear, longitudinally cleft in the furrow of the mandible. *Legs* long, slender, naked above the tarsal joint. *Toes* three before, and one behind; the front ones united at their base by a membrane; that connecting the outer with the middle one always the largest; hind toe short, and barely touching the ground with its tip, or nail; fronts of the tarsi and toes scutellated.

COMMON SANDPIPER.

Totanus Hypoleucos, *Temm.*

Le Chevalier Guinette.

THOSE who have had opportunities of observing this little emigrant in a state of nature must, we think, have been gratified with its tame and inoffensive disposition, and we should suppose would be desirous of knowing more of its history. Unlike many others of its tribe, which are capable of braving with impunity the severities of our hardest winters, the present delicate little bird appears to be adapted to climates of a milder kind, and to inhabit peculiar localities. Arriving here about the end of April, it retreats to inland lakes, rivers, and small brooks, the banks of which it enlivens during the whole of the summer with its active and sprightly habits and simple note. The task of incubation is commenced soon after its arrival, the female depositing her four delicate eggs, of a pale reddish white ground spotted with darker red, on the bank near the water's edge, a mere hollow in the soil or depression in the shingle serving instead of a nest; sometimes, however, it is lined with dried grass, leaves, &c. It is to be feared that the timid disposition of this bird militates much against its security, and that a great portion of those which resort to our navigable rivers and canals fall a sacrifice to the gun, or are otherwise so much disturbed as to prevent their accomplishing this task, for which alone they have been impelled hither; those, on the contrary, which have chosen situations of greater safety and retirement have in a few weeks accomplished the process of incubation, and their half-fledged young soon trip nimbly over the sand and oozy mud, and a short period longer enables them to wing their way after their parents and to seek both food and safety for themselves. Unlike the Dunlin and other marine species whose immense flights almost astonish us, the Common Sandpiper can scarcely be called gregarious, four or six in company being of rare occurrence. Although not very numerous, it inhabits every part of Europe, over which it is dispersed in pairs; and not unfrequently a single individual is observed in solitary retirement, having been bereft of its mate, or by some other circumstance left by itself. In the British Isles the adults commence their autumnal migrations southwards about September, preceding the young by a few weeks, which period allows the latter to gain sufficient strength to perform a fatiguing journey across the Channel to seek retirement in the hotter portions of Europe; and in all probability the northern regions of Africa are not beyond the limits of their annual range. Independently of Europe and Africa, we have observed skins of this species from several parts of India, which proves its range over the Old World to be widely extended. Its place in America is supplied by the well-known Spotted Sandpiper *Totanus macularius*, which occurs in Europe only as a rare visitant.

The sexes are so strictly alike in their colouring as to render a separate description quite unnecessary. The young of the year have the edges of the feathers fringed with a margin of greyish white; in other respects they resemble the adults.

The food of the Common Sandpiper consists of insects of various kinds, to which are added small shelled snails, worms, crustacea, &c., in capturing which, its motions are not less elegant than graceful, running with agility over the oozy mud and sand-banks, often exhibiting a peculiar and singular jerking of the tail and a nodding of the head not unlike that of the Common Water-hen and some of the terrestrial pigeons of the West Indies. Its flight is somewhat slow and flapping, as if performed with considerable exertion, passing so close to the surface of the water as scarcely to avoid wetting the pinions: while flying, it utters its plaintive monotonous call, which is repeated at short intervals until it alights on the opposite side.

Head and upper surface light brown, glossed with olive green; the feathers of the back and scapulars marked with transverse zigzag lines of dark brown, producing a mottled appearance; greater wing-coverts tipped with white; the lesser coverts transversely barred with brown; the first two quill-feathers wholly brown, the remainder of the same colour with a large white spot in the centre of the inner web; the four middle feathers of the tail similar to the back; the two next on each side tipped with white; the outer ones being pale brown with darker bars and a white tip; throat white slightly speckled with pale brown; sides of the neck and breast greyish white streaked with brown; belly white; bill olive; legs and toes greyish yellow.

We have figured an adult and a young bird in autumn, of the natural size.

SPOTTED SANDPIPER.
Totanus macularius. (Temm.)

Drawn from Nature & on Stone by J & E Gould.

Printed by C. Hullmandel.

SPOTTED SANDPIPER.

Totanus macularius, *Temm.*

Le Chevalier perle.

This elegant little Sandpiper is most intimately allied to the well-known Common Sandpiper (*Totanus hypoleucos*), which pays its annual visit during the summer months to the brooks and rivulets of our island ; but, unlike this latter bird, its visits are of the most rare occurrence, no instance having come under our own observation. M. Temminck states that it occurs accidentally on the shores of the Baltic, and in some of the provinces of Germany, but never in Holland. The native country of this bird appears to be the arctic regions of both continents ; but it is most abundant in America, extending from these high latitudes over the whole of the United States, where it appears to take up the same situation as the *Totanus hypoleucos*, frequenting Pennsylvania, and the rivers Schuylkill and Delaware, as we are informed by Wilson, from whose valuable work we have taken the liberty of extracting an account of the habits and manners of this bird, which we have not had the opportunity of observing. " This species is as remarkable for perpetually wagging the tail, as some others are for nodding the head ; for whether running on the ground or on fences, along rails or in the water, this motion seems continual ; even the young, as soon as they are freed from the shell, run about constantly wagging the tail. About the middle of May they resort to the corn-fields to breed, where I have frequently found and examined their nests. One of these now before me, and which was built at the root of a hill of Indian corn, on high ground, is composed wholly of short pieces of dry straw. The eggs are four, of a pale clay or cream colour, marked with large irregular spots of black, and more thinly with others of a paler tint. They are large in proportion to the size of the bird, measuring an inch and a quarter in length, very thick at the great end, and tapering suddenly to the other. The young run about with wonderful speed as soon as they leave the shell, and are then covered with a down of a dull drab colour, marked with a single streak of black down the middle of the back, and with another behind the ear. They have a weak plaintive note." To this we may add, that the young in this stage of its existence, being destitute of the spotted markings of the breast, is very like the young of the Common Sandpiper ; its smaller size, however, will always distinguish it. " On the approach of any person, the parents exhibit symptoms of great distress, counterfeiting lameness, and fluttering along the ground with seeming difficulty. . . . . The flight of this bird is usually low, skimming along the surface of the water, its long wings making a considerable angle downwards from the body, while it utters a rapid cry of *weet, weet, weet*, as it flutters along, seldom steering in a direct line up or down a river, but making a long circuitous sweep, stretching a great way out, and gradually bending in again to the shore."

The tip and upper mandible of the bill " dusky, basal part orange ; stripe over the eye, and lower eyelid, pure white ; whole upper parts a glossy olive, with greenish reflections, each feather marked with waving spots of dark brown ; quills dusky black ; bastard wing bordered and tipped with white ; a spot of white on the middle of the inner vane of each quill-feather, except the first ; secondaries tipped with white ; tail rounded, the six middle feathers greenish olive, the other three on each side white barred with black ; whole lower parts white, beautifully marked with roundish spots of black, small and thick on the throat and breast, larger and thinner and well defined as they descend to the tail ; legs of a yellow clay colour ; claws black."

" The female is as thickly spotted below as the male ; but the young birds of both sexes are pure white below, without any spots : they also want the orange on the bill."

We have figured an adult male, and a young bird of the first autumn, of the natural size.

TURNSTONE.
Strepsilas collaris. (Temm.)

Drawn from Life & on Stone by J & E Gould.

Printed by C Hullmandel.

Genus STREPSILAS.

GEN. CHAR. *Bill* as short as the head, strong, thick at the base, tapering gradually to the point, forming an elongated cone: upper *mandible* the longest, rather blunt at the end. *Nostrils* basal, lateral, linear, pervious, partly covered by a membrane. *Wings* long, pointed, first quill-feather the longest. *Feet* four-toed, three before, one behind, the anterior toes united by a membrane at the base, and furnished with narrow rudimentary interdigital membranes; hind-toe articulated up the tarsus and only touching the ground at the tip.

TURNSTONE.

Strepsilas collaris, *Temm.*

Le Tourne-pierre à collier.

ONLY one species of this genus has hitherto been discovered by naturalists, which, when in the adult state, is as remarkable for the beauty and variety of its plumage as for the singularity of its form. It is found on our shores, and particularly those of our eastern coast, during the greater part of the year, but absents itself from May to July, and proceeds northward to breed. Dr. Fleming states, that having seen this bird at all seasons in Zetland, he concludes it breeds there. It also breeds in Norway, on the shores of the Baltic, in the North Georgian Isles; and our intrepid Arctic voyagers found it at Melville Island, from whence they brought some of its eggs. This bird is found also in Africa and America.

The Turnstone frequents the sandy and gravelly parts of the sea-shore, where it feeds upon insects, small mollusca, and crustacea, which it finds under stones, for the turning over of which its wedge-shaped beak is admirably adapted. In its habits it differs from the Sandpipers generally, as it is not observed to fly in flocks, or, like them, to frequent the soft and oozy mud left by the retiring tide. It is lively and quick in its motions, and runs from place to place in search of its food with rapidity.

The adult male has the forehead, the space between the beak and the eye, throat, nape and side of the neck, lower part of the back, upper tail-coverts, breast, and all the under parts, pure white; the top of the head is mottled with black; below the eye and on the sides and front of the neck the plumage is black, with two narrow black bands passing backwards from the upper and under edges of the base of the beak. The black feathers on the side and bottom of the neck also extend backwards, forming two collars more or less perfect; the back, scapulars and wing-coverts, are reddish brown varied with black; primaries black on the outer webs, secondaries tipped with white; rump black; outer tail-feathers on each side white, the others black tipped with white; beak black; irides dark drown; legs orange red.

The plumage of the female is generally less brilliant than that of the adult male.

Young birds of the year have the throat white; the darker parts of the head and neck ash brown; back, scapulars and wing-coverts dusky brown, with lighter edges; all the under parts white. As the season advances, the feathers on the lower part of the neck in front become nearly black; the centre of the feathers on the back and wings much darker, with broad rufous edges, assuming by degrees the brilliancy of old birds, which is nearly acquired by the end of the following spring.

Some difference of opinion exists as to the colour of the eggs of the Turnstone. By M. Temminck and some other foreign naturalists it has been described and figured as having a green or ash-coloured ground, spotted with dark brown or black. Mr. Lewin's figure represents it ash green with spots of two colours, both dark. An egg, marked "Turnstone," in the extensive collection at the British Museum has a reddish white ground spotted with dark red, and we have seen one of the specimens brought from Melville Island which was exactly of the same colour.

We have figured a male in his adult plumage, and a young bird in that of the first autumn.

WOODCOCK.
Scolopax rusticola, /Linn./

Drawn from Nature & on stone by J.&E.Gould.

Printed by C.Hullmandel.

Genus SCOLOPAX, *Linn.*

Gen. Char. *Bill* long, straight; the tip obtuse, rounded, and ending with an internal knob; both mandibles, in dead birds, rugose behind the tip; under mandible shorter than the upper, which is sulcated for nearly the whole of its length. *Nostrils* basal, lateral, placed in the commencement of the groove, linear, longitudinal, covered with a membrane. *Wings* having the first and second quills of nearly equal length and the longest. *Legs* slender; the tibiæ either entirely plumed or naked for a short space above the tarsal joint. *Feet* four-toed, three before and one behind, the former cleft to the origin, the latter short and its tip only resting on the ground.

WOODCOCK.

Scolopax rusticola, *Linn.*

Le Becasse ordinaire.

So well has the history of this familiar, and to the sportsman favourite, bird been detailed by various British and Continental authors, among whom we may especially mention Mr. Selby, that we shall confine our remarks more to its geographical distribution than to those minor details with which most persons must be familar: but before entering upon these particulars, we would here express our decided opinion that the present bird, with the Woodcock of the United States of America, and, if we mistake not, one or two other species, may with strict propriety be separated into a distinct genus, which has indeed been done by M. Vieillot under the name of *Rusticola*; for not only is the difference in form between the Woodcock and Snipe very apparent, but there is a still greater diversity in their habits and manners.

In England and we believe in nearly every portion of Europe, the Woodcock is a migratory species: a few pairs, it is true, stay with us to breed, but the great mass undoubtedly pass northwards, even to within the limits of the arctic circle, tenanting during summer the wilds and forests of that desolate region; and as soon as the work of incubation is over, they commence in vast hordes their southward flight, our island being merely a resting-place for a large portion of them in their progress towards still more southern latitudes; hence the promontories of Ireland, Wales, Cornwall, and Devonshire abound with them at the periods of their vernal and autumnal migrations. From these summer haunts, which appear to extend throughout the Old World portion of the zone, they radiate southwards, not only through Europe, but to the vast regions of Asia, penetrating even into India, whence we have received numerous specimens differing in no respect from those killed in our own island.

There is scarcely any difference in the appearance of the sexes, and they do not undergo any decided periodical change in their plumage.

The nest is generally placed in a thicket, near the root of a tree or shrub, and is merely a slight hole lined with a few dead leaves and grass; the eggs are four in number, of a yellowish white, blotched with pale chestnut brown, and in Sweden and other parts of the Continent are considered a great delicacy for the table.

The food of the Woodcock consists principally of worms, which it procures by inserting its long bill into the earth.

Forehead and top of the head grey; hind part of the head and neck marked with four broad brownish black bars; the intermediate spaces reddish white; from the gape to the eye a streak of deep brown; chin white; on each side of the neck a patch of brown; upper surface a mixture of rufous brown, pale dull yellow and grey, with large spots and zigzag transverse lines and bars of black, which colour is deepest on the back and scapulars; rump and tail-coverts pale chestnut brown with pale reddish white tips and narrow transverse bars of black; tail black varied with chestnut brown; the tips of the feathers grey above and pure white below; quills dusky, outer webs having triangular bars of chestnut brown; under surface greyish white tinged with yellowish brown and barred transversely with brown of a darker tint; vent and under tail-coverts yellowish white, with a triangular spot of black in the centre; legs flesh-red tinged with grey.

We have figured an adult bird of the natural size.

GREAT SNIPE.
Scolopax major. (Gmel.)

Drawn from Nature & on Stone by J & E Gould.

Printed by C. Hullmandel.

GREAT SNIPE.

Scolopax major, *Gmel.*

La Grande ou Double Bécassine.

THE specific appellation given to this bird, as is the case in many other instances, shows the impropriety of such names as *major, minor, minuta,* &c. ; for although we admit that the present bird is the largest of the European true Snipes, still there are two others which exceed it in size from the hilly districts of India, and a third from Mexico, whose size is even superior to that of the Woodcock : the name of *major* as applied to our bird is therefore perfectly inappropriate.

The northern parts of Europe undoubtedly constitute the true habitat of the Great Snipe. Sir Humphry Davy killed several during one of his summer visits to Norway, &c. : these were afterwards transmitted to the Zoological Society of London, and on examination were ascertained to be strictly identical with those killed in England. We do not mean to affirm that Norway and the northern regions generally are its sole habitat, for we have received it in abundance from the temperate and southern portions of Europe and the borders of Asia. Although we have no direct evidence that it breeds in the British Islands, still it is far from being improbable that instances of the kind may occasionally take place, particularly as it appears to be more common than has been hitherto supposed.

The term *Solitary*, by which it is known in some parts of England, is not inappropriately applied to this Snipe, in as much as it is always found alone, and, as it were, isolated from the companionship of others of its species ; neither does it appear to congregate into bodies for the purpose of migration, each individual, or at most each pair, seeming to act independently for itself.

In its general appearance the Great Snipe closely resembles the Common Snipe (*Scolopax Gallinago*), but on minute examination the flanks will be found to be strongly barred with brown, the secondaries and lesser wing-coverts numerously spotted with white, and the outer tail-feathers totally destitute of any markings ; in addition to which it is much heavier, the weight of the Common Snipe being rarely more than four ounces, while the Great Snipe frequently weighs seven or eight. Its flight is less tortuous and rapid than that of the Common Snipe, being performed in a more steady and even line, and is not unlike that of the Woodcock.

Its food consists of insects, which it procures by thrusting its bill into the soft and oozy mud.

Like its congeners it is principally found in heaths, low marshy situations, morasses, &c.

Its habits, nidification, &c. are said to be precisely the same as those of the Common Snipe, and its flesh is equally esteemed as a delicacy for the table.

Crown of the head dark brown, interspersed with small markings of reddish brown, with the central streak of the same colour ; a streak of pale buff between the bill and the eye ; back dark brown varied by longitudinal markings of yellowish brown ; lesser wing-coverts tipped with white ; breast, sides, and flanks white with transverse triangular bars of deep brown ; tail of sixteen feathers, the two centre ones black for two thirds of their length ; the outer feathers quite white for nearly their whole length ; legs olive ; bill blackish brown.

The Plate represents an adult male of the natural size.

Drawn from Nature & on stone by J.&E. Gould.

1. SABINE'S SNIPE.
Scolopax Sabini. (Vigors).

2. COMMON SNIPE.
Scolopax Gallinago. (Linn.)

Printed by C.Hullmandel.

SABINE'S SNIPE.

Scolopax Sabini, *Vigors.*

THE occasional occurrence of this rare and singular species of Snipe in our island, teaches us that we have yet much to learn respecting the native localities of many of the feathered tribes, for we know of no instance of its having been killed in any other part of the globe than the British Islands; still it is very evident that these islands are not its native home, and that those that have been killed here are merely stragglers from some unknown region. The first example of this bird was killed in Queen's County, Ireland, in August 1822, and was sent to Mr. Vigors the same day; it was described by him under the above title in the 14th vol. of the Transactions of the Linnean Society, and is now contained in the Museum of the Zoological Society, to which institution it was presented by Mr. Vigors with the whole of his fine collection. A second example was shot on the banks of the Medway, near Rochester, in October, 1824. Besides these, Mr. Selby informs us that he has "received a fresh specimen of this rare Snipe from Morpeth, possessing all the characteristics of Mr. Vigors's bird;" and we ourselves know of another example having been killed in Ireland.

As we are indebted to Mr. Vigors for our knowledge of this species, we deem it but just to quote his own words in pointing out its distinctive characters. "It is at once distinguished from every other European species of *Scolopax* by the total absence of white from its plumage, or any of those lighter tints of ferruginous yellow which extend more or less in stripes along the head and back of them all. In this respect it exhibits a strong resemblance to *Scolopax saturata* of Dr. Horsfield, from which, however, it sufficiently differs in its general proportions; and I find no description of any other extra-European species of true *Scolopax* which at all approaches it in this character of its plumage. In the number of its tail-feathers, again, which amount to *twelve*, it differs from *Scolopax major*, which has sixteen, and *Scolopax Gallinago*, which has fourteen; it agrees, however, in this point with *Scolopax Gallinula*, which also has but twelve; but it can never be confounded with that bird from the great disproportion between the essential characters of both, the bill alone of *Scolopax Sabini* exceeding that of the latter species by one third of its length. In the relative length and strength of the tarsi it equally differs from all. These members, although stouter than those of *Scolopax Gallinago*, fall short of them by $\frac{7}{10}$ths of an inch; they are much weaker, on the other hand, than those of *Scolopax major*, although they nearly equal them in length." Of its habits, mode of nidification, &c., we know nothing; but in these respects it doubtless bears a close resemblance to the other members of the genus.

Top of the head and back black, the latter being transversely barred with chestnut; whole of the under surface dusky black, thickly barred with dull chestnut; quills blackish brown; tail of twelve feathers, black at the base, chestnut at the tip barred with narrow lines of black; bill dusky black, the base of the upper mandible pale chestnut; legs dark olive green.

COMMON SNIPE.

Scolopax Gallinago, *Linn.*

Le Bécassine ordinaire.

ALTHOUGH the contrary has been long recorded by naturalists, we conceive that the natural range of the Common Snipe is comparatively limited, and that the Snipes from India, Africa, and North America, that have been regarded as identical with our bird, will be found, on examination, to be specifically distinct; in the character of their plumage they are indeed somewhat similar, but they nearly all present a different form in the feathers of the tail, and also a difference of number.

The Common Snipe is strictly indigenous in our islands, although the great mass retire northwards to breed, leaving a few scattered over our extensive moors and marshy districts, where they perform the task of incubation: these few have their numbers augmented in autumn by the return of those which had retired to northern latitudes, whence they are now driven with their young by the severity of the climate and the impossibility of acquiring food. On the Continent it inhabits the same situations as in Great Britain and is equally abundant. Its habits, manners, mode of life, and flight are so universally known that a detailed account of them is perfectly unnecessary, neither need we say anything about the excellency of its flesh as an article of food. The nest is usually formed by lining some small depression of the ground, under a tuft of grass, heath, or rushes, with dried grasses and similar materials; the eggs are four in number, long and pointed, of an olive green blotched with different shades of reddish brown. The young quit the nest almost immediately after their exclusion from the shell, and run nimbly about after their parents while yet covered with a particoloured dress of brown and buff. The sexes offer so little difference in the markings of their plumage, that it is impossible to distinguish them by this means.

In the adult bird the top of the head is brown, divided by a longitudinal central stripe of yellowish white; a similar stripe of yellowish white runs from the base of the beak above the eye, followed by a stripe of brown from the base of the beak to the eye; chin white; sides of the neck and chest tawny yellowish white, numerously spotted with dark brown; back and scapulars fine black barred with brown, and with longitudinal stripes of rich buff yellow on the outer edges of the feathers; wings dark brown, each feather being spotted and edged near the tip with yellow; primaries dark brown, the outer web of the outer quill being white; under surface white, barred on the flanks with brown; tail consisting of fourteen feathers which are black for two thirds of their length from the base, the rest reddish brown with a bar of black and tipped with reddish brown; legs and feet greyish olive; beak yellowish brown becoming redder at the base and darker at the tip.

We have figured a male of both species of the natural size.

JACK SNIPE.

Scolopax Gallinula (Linn.)

JACK SNIPE.

Scolopax Gallinula, *Linn.*

Le Becassine sourde.

THE Jack Snipe, although equally as abundant as the Common Snipe during the autumn and winter, quits us entirely on the approach of spring, and retires to more northern countries, probably within the regions of the arctic circle, where, in company with numerous others of the feathered race, it remains to incubate, and again returns to us in the months of October and November, when as long as the weather is open it may be found in any of the marshy districts of this country, and throughout the Continent generally. We have reason to believe, however, that Europe alone constitutes its true and almost exclusive habitat; for although it may be occasionally met with out of Europe, it is extremely rare. Among all the numerous collections from the Himalaya mountains we do not recollect having seen more than one specimen. It is by far the least of the true *Scolopacidæ*, its weight being seldom more than two ounces. The Jack Snipe usually frequents the same localities as the Common Snipe, but differs from it considerably in its habits and manners; for while the latter is somewhat shy and easily flushed, the Jack Snipe, on the contrary, will frequently allow itself to be almost trodden upon before it can be forced to take wing. And we cannot fail to remark how beautifully the colouring of this bird assimilates with the ground and the surrounding herbage among which it lies, which, together with its motionless manner of lying, renders it most difficult to be discerned, unless the spot on which it sits is most carefully and scrutinously examined. The individual from which our drawing is taken was captured alive by ourselves with the hand, from before the nose of the pointer.

Its flight, although often extremely rapid, is seldom prolonged to any distance, the bird generally alighting again immediately, except on the approach of the vernal migration, when we have seen it mount in the air and totally disappear, without even uttering the alarm cry usual with this and other members of the genus. Its flesh has a most delicate flavour, but from its diminutive size it is not so much sought after by the sportsman as the larger common species.

It is said to breed in bogs and morasses, and according to M. Temminck the eggs are four in number.

The sexes offer no difference in the markings of their plumage, which undergoes little or no change in spring or summer. The young acquire the adult colouring, although not so bright, from the time they leave the nest.

A band of black spotted with yellowish red extends from the forehead to the nape; a distinct band of buff passes over the eye; the remainder of the face is alternately striped with black and light buff; throat whitish; upper part of the chest yellowish brown, blotched with spots of brown; back and scapulars blackish brown with green and bronze-like purple reflections; the latter feathers are long and narrow, and have their outer edges of rich buff, forming two longitudinal bands down each side; wing-coverts blackish brown, each feather margined with light brown; tail brownish black edged with rich brown; abdomen white; flanks and lower parts longitudinally streaked with brown; legs olive; bill greyish olive.

The Plate represents two birds of the natural size.

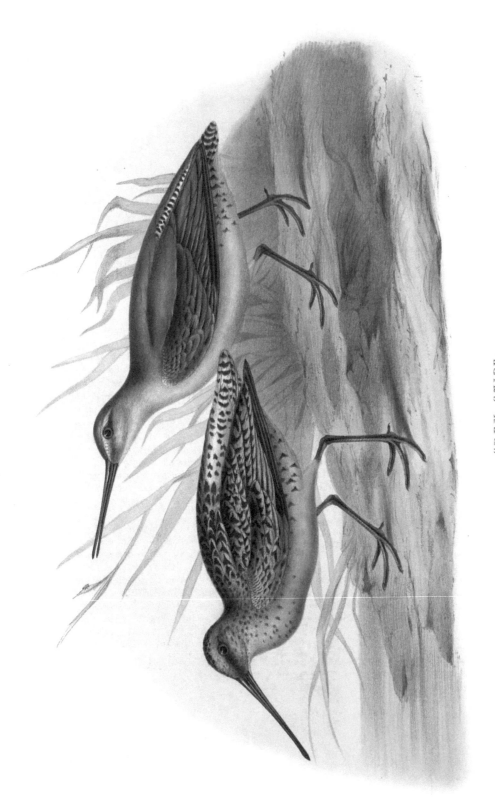

GREY SNIPE.

Macroramphus Griseus (Leach.)

Drawn from life & on Stone by J. & E. Gould.

Printed by Hullmandel

Genus MACRORHAMPHUS, *Leach.*

GEN. CHAR. *Beak* very long, straight, rounded, rather slender in the middle, the tip dilated, somewhat incurved and rugose. *Nostrils* lateral, basal. *Legs* four-toed; the outer *toes* connected at their base by a membrane; hinder *toe* touching the ground only at the tip; lower part of the tibiæ naked. *Wings* long and pointed.

GREY SNIPE.

Macrorhamphus griseus, *Leach.*

La Becassine ponctuée.

THE bird now before us has the beak of a *Scolopax*, but with this it also possesses toes connected by a membrane; to this we may add, that its habits and the peculiar periodical change of plumage to which it is subject, still further prove its alliance to the species of the genus *Tringa*. Dr. Leach, however, considered it sufficiently removed both from Snipes and Sandpipers to warrant the adoption of a separate generic distinction, and we have followed his example.

This bird occurs but seldom in Europe. A single specimen has been taken in Sweden. A second was obtained in England on the Devonshire coast, which came into the possession of Colonel Montagu, and is preserved in the British Museum : this specimen is in the plumage peculiar to the winter season. A third example has been shot at Yarmouth, which was in the plumage of summer. No other instances, that we are aware of, are recorded. It is, on the contrary, very plentiful on the Western shores of North America, from the United States, even to the Arctic circle; and to the excellent work of Wilson we are indebted for the best account of its habits and œconomy.

" The Red-breasted Snipe," as it is called by Wilson, on account of the prevailing colour of its summer plumage, " arrives on the sea-coast of New Jersey early in April; it is seldom or never seen inland : early in May it proceeds to the North to breed, and returns by the latter part of July or beginning of August. During its stay, it flies in flocks, sometimes very high, and has then a loud and shrill whistle, making many evolutions over the marshes; forming, dividing, and reuniting. They sometimes settle in such numbers, and so close together, that eighty-five have been shot at one discharge of a musket. They frequent the sand-bars and mud-flats at low water, in search of food; and being less suspicious of a boat than of a person on shore, are easily approached by this medium, and shot down in great numbers."

" These birds," says Wilson, " of all our sea-side Snipes, are the most numerous, and the most delicious for the table. They doubtless breed not far to the northward of the United States, if we may judge from the lateness of the season when they leave us in spring, the large size of the eggs in the ovaries of the females before they depart, and the short period of time they are absent." This hiatus in their history has been supplied by Dr. Richardson in his Fauna of North America, who says, " they are well known in the fur-countries, and have an extensive breeding range from the borders of Lake Superior to the Arctic Sea.

The whole length of this bird is nearly eleven inches, the bill two inches and a half. In winter, the head, neck, breast and wing-coverts are uniform ash-coloured brown; a streak of the same colour from the base of the beak to the eye; throat, belly and thighs pure white; flanks white, varied with light brown; back and scapulars light brown, each feather with a darker brown edge; rump, upper and under tail-coverts white, with black cross bars; tail-feathers twelve, crossed with narrow black and white bars alternately; the beak dark greenish black; legs dull green. In summer, the top of the head and neck, back and scapulars are irregularly varied with black, reddish brown, and yellow; the space round the eye bright red; front of the neck and breast reddish buff; wing-coverts ash-colour and edged with white; belly, rump and tail the same as in winter. The female is paler on the back, and less ruddy on the breast.

The figures on our Plate represent the plumage of both seasons.

KNOT.
Calidris canutus.

Genus CALIDRIS.

Gen. Char. *Beak* mediocral, rounded, rather slender, flexible, above sulcated, its tip depressed, smooth and dilated. *Nostrils* linear, placed in a long groove. *Feet* four-toed; the toes margined and cleft; the hinder toe with its tip alone resting on the ground.

KNOT.

Calidris canutus, *Briss. et Cuv.*

Le Bécasseau canute.

Although the Knot does not make the British Islands a place of permanent residence in which to incubate and rear its young, yet it is abundant on its passage both during its equinoctial and polar migration. While on the latter journey, it visits our island in the month of May, at which time many individuals have almost entirely gained the full plumage of summer, so much so, as to lead us to suppose that they would remain on our shores to breed; a supposition which is strengthened by actual dissection; still, however, we have no undoubted facts upon which to assert that such a circumstance has occurred. Leaving our shores after a short sojourn, they pass northwards to their arctic breeding-places: this duty accomplished, they commence their migration southward, visiting our island again in their passage, when many remain the whole of the winter, during which they live congregated in flocks on the borders of the sea, but giving the preference to marshy and fenny countries in their native latitudes for the purpose of incubation. While they make this island their asylum, numbers are annually taken, either in nets or by the gun, for the purposes of the table. In the London markets they may be generally met with in the spring, and during the whole of the winter.

The Knot is not only common in the arctic regions of the Old World, but is equally so in the northern portions of America, extending throughout the whole of the circle; its southern migrations seldom exceeding the latitudes of the Mediterranean,

We know of no birds in which the great difference between the stages of plumage in winter and summer has led to so much confusion and the creation of so many synonyms: an examination of individuals in every stage of plumage, from the greyish white of winter to the fine brownish red of summer, has clearly satisfied us that they are specifically the same. Independently of the great dissimilarity of colour which the adult Knot exhibits at these opposite seasons of the year, the young possess a colouring distinct from either, the ground of which nearly resembles the plumage of winter, but every feather on the upper surface is edged at its tip with two crescents, the outer one white and the inner one black, producing a most beautifully barred appearance: under surface buff colour. Concerning the nidification of the Knot we have been able to collect no information whatever.

The two sexes are alike in colouring, or, if there be any difference, the female is the finest in colour and the largest in size.

In summer, the whole of the upper surface is of a reddish brown, the top of the head and back of the neck being marked with small longitudinal streaks of black, while each feather on the back and wings has a central dash of the same colour branching out into irregular bars on the wings; quill-feathers and tail blackish brown; the whole of the under surface is of a brick red; bill green at the base, black at the tip; legs greenish olive; irides dark brown.

In winter the whole of the upper surface is of a fine ash grey; the quills and tail brownish black; the breast and whole of the under surface white, flanks and sides of the chest being variegated with longitudinal and arrow-shaped marks of brown.

The Plate represents an adult bird in the summer and winter states of plumage of the natural size.

RUFF.

Machetes pugnax. *(Linn.)*

Drawn from Nature & on Stone by J.E. Gould.

Printed by C. Hullmandel.

Genus MACHETES, *Cuv.*

Gen. Char. *Bill* straight, rather slender, as long as the head, with the tip dilated and smooth. *Nostrils* basal, lateral, linear, placed in the commencement of the groove. *Wings* long and sharp-pointed; first and second quill-feathers equal and longest. *Legs* long and slender. *Tibiæ* naked for a considerable space above the tarsal joint. *Feet* four-toed; three before, and one behind; the outer toe united to the middle one by a membrane as far as the first joint; hind toe short.

The head and neck of the male, during the breeding-season, are adorned with long plumose feathers, springing from the occiput and throat, which, when raised, form a large ruff around the head; and the face of the male bird, during the same period, is covered with small fleshy warts or papillæ.

RUFF.

Machetes pugnax, *Cuv.*

Le Bécasseau combattant.

The species of the great family of Sandpipers, from which this remarkable bird has been separated into a distinct genus, are well known to undergo a striking periodical change of plumage immediately preceding the season of reproduction; but it is in the present bird alone that we find so great a diversity of colour in the plumage during this period, as to render it scarcely possible to discover two individuals exactly according with each other either in tints or markings. In one, for example, we find the frill of a beautiful buff, with elegant bars of black; in another, white, grey, or chestnut, with longitudinal markings of shining black with violet reflections. Others, again, have the frill of a uniform white, black, or brown, with auricular feathers of a different colour. The remainder of the plumage (and it is even the same with the colouring of the legs and bill) undergoes a considerable change, though not to so remarkable a degree as do the feathers of the neck. With the accession of these ornamental plumes, the ruff also acquires a multitude of small warty tubercles, thickly disposed over the front part of the head. To figure the whole of these spring changes common to the male is of course impossible; we have therefore selected for our Plate an example the colour of which could be most faithfully represented. This remarkable change, by which the Ruff is so much distinguished, commences and is completed during the month of April. In this, which we may call its most perfect state of plumage, it continues about a month, when the plumes are gradually thrown off, and by the end of June it assumes its more common appearance, which it retains during the remainder of the year, no difference being then to be perceived between the individual whose frill had been white, and the one in which it had been black. The females, or Reeves, are much smaller than the male, are never adorned with the elongated feathers of the neck, and vary but little in their plumage throughout the year, which is very similar to that of the male in his winter dress.

In England the Ruff is not so abundant as it was formerly: its chief resorts now are the fens of Cambridgeshire, Lincolnshire, and Norfolk, although some few are occasionally found in other places. In Holland and the low and marshy districts of France and Germany it is in the greatest abundance. From these countries it spreads northwards to very high latitudes during the summer, as it is known to breed in Lapland, Sweden, and Russia.

The Ruff may be considered a strictly migratory species, retiring to southern latitudes in winter, and arriving in our own in the early part of spring, the males making their appearance several days previous to their expected mates. Unlike the *Tringæ* in general, the Ruff is polygamous in its habits, each male claiming to himself a certain district, the limits of which he defends with the utmost obstinacy. On the arrival of the females, the males pay their attentions by various displays of their plumage, which is now in full perfection; and as many lay claim to the same female, sharp contests ensue for the possession of her, and she becomes the prize of the conqueror.

On account of the great delicacy of its flesh, the Ruff is highly esteemed for the table; hence various means are resorted to for the purpose of securing them alive, in order that they may be fattened in confinement. They soon become extremely tame and familiar; are fed with bread and milk, boiled wheat and other farinaceous seeds, which they eat with great avidity; and speedily become very fat. In a state of nature their food consists of worms, aquatic insects, and their larvæ.

As the description of one male in his summer dress will not be applicable to any other individual, we shall content ourselves with giving that of the male in winter, which is the same in every example.

The whole of the upper surface brown, each feather having its centre of a deeper tint, and its margin of a light reddish grey; the whole of the under surface is white; feet and bill light brown.

The female, or Reeve, is full a third less than the male, and closely accords with him in his winter dress, except that the throat, fore part of the neck, and breast are light brown, mingled with darker blotches.

The Plate represents a male in summer and winter plumage, and an adult female, all of the natural size.

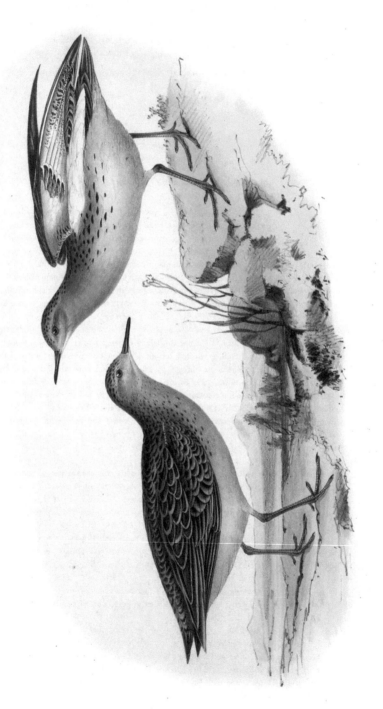

BUFF-BREASTED TRINGA.
Tringa rufescens. (Vieill.)

Drawn from Nature & on stone by J. & E. Gould.

Printed by C. Hullmandel.

BUFF-BREASTED SANDPIPER.

Tringa rufescens, *Vieill.*

Le Tringa roussâtre.

THIS prettily marked Sandpiper was first made known (Linn. Trans., vol. xvi.) as an occasional visitor to England by Mr. Yarrell, who possesses a specimen killed in Cambridgeshire in 1826. In the following year a specimen was killed in France, which is now deposited in the Paris Museum; since which another example has been killed in England at Sherringham in Norfolk, which is preserved in a collection of birds in that county.

This species was first discovered in Louisiana by M. Vieillot, and was unknown to Wilson and other American ornithologists of that time. Mr. Thomas Nuttall of Boston, the author of 'A Manual of the Ornithology of the United States and Canada,' says this elegant species in some seasons is not uncommon in the market of Boston in the months of August and September, being met with near the capes of Massachusetts Bay. It has also been obtained in the vicinity of New York. From its having been found in Louisiana, Mr. Nuttall infers that coursing along the shores of the Mississippi, and thus penetrating inland, it probably proceeds by this route, as well as by that of the sea coast, to its northern destination to breed, and is often seen associated with the Pectoral Sandpiper, which it resembles very much in size and bill, though perfectly distinct in plumage.

Mr. Audubon appears not to have met with this bird in America, beyond procuring a specimen at Boston. The geographical range of this Sandpiper is very extensive. Captain James Clark Ross possesses a wing of this species, (which from the very peculiar markings of its under surface cannot be mistaken,) received from a sailor of the crew, who found it in the course of one of the numerous inland excursions in the desolate regions of the North from which these intrepid navigators have recently returned. It is therefore probable that it breeds near the arctic circle. From M. Natterer we learn that this species is common in Brazil, and we have already noticed its occurrence three times in Europe.

Mr. Nuttall states that in America its food consists principally of land and marine insects, particularly grasshoppers, which, abounding there in autumn, become the favourite prey of a variety of birds.

Top of the head dark brown, the feathers edged with very light brown; back of the neck light brown, with minute longitudinal darker spots; the back dark brown, the extreme edges light brown; wing-coverts brown; primaries nearly black, tipped with white, the shafts white; tertials brown, edged with light brown; tail cuneiform; the coverts brown, with lighter borders; the centre feathers black; the shafts and edges lighter; the feathers on each side light brown, inclosed by a zone of black and edged with white; chin, sides of the neck, throat, and breast light brown, tinged with buff; abdomen, flanks, and under tail-coverts white; sides of the neck spotted; anterior portion of the under wing rufous brown; under wing-coverts pure white; inner webs of the primaries speckled; secondaries mottled, and ending in sabre-shaped points; the legs light brown.

Mr. Audubon informs us that the female is somewhat larger than the male, which it resembles in colour, but has the lower parts paler, and the feathers of the upper parts of a lighter brown, with an inner margin of brownish black and an outer one of greyish yellow.

In young birds the tints are said to be much lighter, the primaries more spotted, some of the inner wing-coverts also mottled; all the upper plumage more broadly edged with pale buff, on the back inclining to white. The colour beneath is also buff, becoming almost white on the belly and vent.

We have figured an adult male and female of the natural size.

PECTORAL SANDPIPER.
Tringa pectoralis, *(Bonap.)*

PECTORAL SANDPIPER.

Tringa pectoralis, *Bonap.*

Le Becasseau Pectorale.

An example of this species of Sandpiper having been killed in our island, we have deemed it necessary to include a figure of it in the present work. As we have nothing to add to the account published by Mr. Hoy in Loudon's Magazine of Natural History, we think it best to quote that gentleman's words: "The occurrence of the Pectoral Sandpiper, *Tringa pectoralis*, is noticed and a plate given by Mr. Eyton in his continuation of Bewick's *Birds*. I am not aware of a more recent instance of its occurrence, and have thought it might be interesting to some of your readers to know something more respecting the capture of the above-named specimen. This *Tringa* seems allied both to *T. variabilis* and *T. subarquata*; and in the form of the bill shows some affinity with the Knot (*T. canutus*). In size it is superior to the Curlew Tringa (*T. subarquata*). It was killed on October 17th, 1830, on the borders of Breydon Broad, an extensive sheet of water near Yarmouth, rather celebrated for the numerous rare birds which have at different times been observed and shot on its banks and waters. The person who killed it remarked that it was solitary, and its note was new to him, which induced him to shoot it. It proved a female on dissection.

"This specimen has been examined by M. Audubon, and identified by him with the *Tringa pectoralis* of North America," which is its true habitat.

Crown of the head, all the upper surface, wings and central tail-feathers dark blackish brown, which is bounded with ferruginous and margined with cinereous; stripe over the eye, chin, abdomen and under tail-coverts white; sides of the face, back and sides of the neck, and the breast pale brown with a stripe of dark brown down the centre of each feather; bill reddish yellow at the base, black at the tip; feet greenish yellow.

Our figure is of the natural size.

PIGMY CURLEW.
Tringa subarquata (Temm.)

Drawn from life & on Stone by J. & E. Gould.

Printed by C. Hullmandel.

PYGMY CURLEW.

Tringa subarquata, *Temm.*

Le Bécasseau cocorli.

THE Pygmy Curlew in its winter plumage has been frequently confounded with the Purre; the beak, however, is longer, rather more slender, as well as more curved; the legs longer and thinner, and the bare part above the joint of greater extent; there is also a constant and marked difference in the upper tail-coverts, which in this bird are invariably white, but in the Purre the central tail-coverts are of the same colour as the feathers of the back. In their decided summer plumage and the various consequent vernal and autumnal changes, in both, the differences are too obvious to require particular notice.

The Pygmy Curlew has been considered also a very rare British bird, and one that did not breed in this country: we have reason, however, to believe,—from the various specimens we have seen and obtained in their most perfect nuptial dress, some of which will be more particularly referred to hereafter, and the several young birds which could only very recently have quitted their nest,—that the Pygmy Curlew breeds every year on various parts of our coast. We have ourselves shot the male, in full summer plumage, at the end of May in the present year (1833), near Sandwich, and have received adult birds equally fine, with the young, from Yarmouth early in July. In its habits it resembles the Purre, flying in flocks in company with other shore birds, and like them also feeding on marine insects, worms, minute mollusca, and crustacea. The male specimen, in summer plumage, from which our right-hand figure was drawn and coloured, we killed out of a flock, and brought down at the same shot a Purre and a Ring Dottrell.

The Pygmy Curlew frequents the shores of the European continent generally, being most observable in spring and autumn: it also inhabits Africa and North America. In its winter plumage, represented by the bird on the left side of our Plate, the throat, neck, breast, all the under parts, and the upper tail-coverts, are pure white; crown and sides of the head, back, scapulars and wing-coverts, ash brown, the shaft and middle of each feather being rather darker; wing-primaries black; tail-feathers ash colour edged with white; beak black; irides dark brown; legs brownish black. During the season of producing the young, the feathers on the top of the head are varied with spots of black and reddish brown; throat, breast and abdomen chestnut red, some of the feathers tipped with white; upper and under tail-coverts white slightly barred across with black and red; back, scapulars and tertials nearly black, the feathers varied on their margins with red and ash grey; some of the wing-coverts remain unchanged; the primaries black; tail-feathers dusky brown with lighter edges.

The intermediate states of plumage, as they appear in spring and autumn, may be inferred from a previous knowledge of the appearance of the bird in winter and summer, the feathers on the breast changing by degrees from white to red, and afterwards regaining the white; those on the back alternating between ash colour and red brown.

The young birds of the year most resemble the adult bird in winter; but the feathers on the upper surface of the body and wings have broad edges of yellowish white; the under surface tinged with buff colour; the legs brown.

M. Temminck states, that this bird occasionally breeds in Holland near the edge of the water, laying four eggs, yellowish white, spotted with brown.

We have figured two birds of their natural size.

DUNLIN, OR PURRE.

Tringa variabilis. *(Meyer)*

Drawn from Nature & on stone by J & E Gould.

Printed by C Hullmandel.

DUNLIN, OR PURRE.

Tringa variabilis, *Meyer*.

Le Becasseau brunette, ou variable.

In consequence of the remarkable changes to which this *Tringa* is subjected, it has in its various stages received several specific appellations, therefore to the one now generally adopted, *variabilis*, is attached a great number of synonyms.

In its winter or grey dress it is called the Purre, and it is at this season that it is most plentifully distributed along the whole line of our coast, where it may be observed congregated in vast flocks, enlivening the bleak and dreary beach by the celerity with which it runs over the sands, and by its sweeping and vigorous flight, during which, like many other of its congeners, every individual of the flock, be it ever so numerous, simultaneously exposes the upper or under surface of the body, as they sweep along over the surface of the ocean or across the sands.

On the approach of spring, the great mass which have wintered in the British Islands retire northwards to breed. At this period a strongly contrasted change takes place in the colouring of the plumage, the uniform grey of winter being superseded by the more rich colouring, which is represented on our Plate. It is in this latter state that it is known by the name of Dunlin.

When the breeding-season commences the greater portion of these birds leave the sea-shore, and retire inland to wild heaths and the upland country, availing themselves of every situation on their passage that affords a suitable retreat for the rearing of their young. They generally select similar spots to those chosen by the Common Snipe (*Scolopax Gallinago*, Linn.), to which bird it assimilates in the whole process of incubation.

Although the greater number of the Dunlins annually migrate northwards, a few solitary pairs always remain in the northern portions of England and Scotland; in the Orkney and Shetland Islands they are still more abundant, and their numbers gradually increase as we proceed further north, until we arrive at the Arctic Circle.

On the Continent the Dunlin is as abundant as it is with us, being universally dispersed and subjected to the same natural laws. We would here also mention, that although the sea-coast constitutes their principal place of residence, they are also found on the banks of inlets and streams, as well as on those of the larger rivers, both on the Continent and in our islands.

The nest is merely a depression in the ground, lined with a few straws or dried grasses: the eggs, four in number, are of a greenish grey, spotted all over with reddish brown.

The food consists of worms, insects, mollusca, and the small crustacea, which it obtains by following the ebb-tide.

The great changes which the Dunlin undergoes rendering it necessary to describe the summer, winter, and immature plumage, we take the liberty of availing ourselves of the accurate description published by Mr. Selby, who has paid great attention to the subject.

"Winter plumage. Crown of the head, hind part of the neck, back, and scapulars ash grey, with a tinge of hair brown, the shaft of each feather being darker; between the bill and the eye an indistinct line of brown; eye-streak and cheeks white, streaked with pale hair-brown; chin and throat white; lower part of the neck and breast grey, the shafts of the feathers hair-brown; under surface pure white; wing-coverts hair-brown margined with pale ash grey, the larger ones having white tips; rump and upper tail-coverts deep brown margined with paler; two middle tail-feathers deep brown, the rest on each side grey, with white shafts; bill black; legs and toes blackish grey.

"Summer plumage. Crown of the head black, each feather margined with reddish brown; chin white; cheeks, fore part of the neck and breast black, with the feathers deeply margined with white, giving these parts a beautifully spotted appearance; under surface black; flanks and side-coverts of the tail white, streaked with black; back part of the neck, mantle, and scapulars black, each feather deeply margined with clear reddish brown; lower part of the back and upper tail-coverts brownish black; wing-coverts as in the winter plumage.

"Immature plumage. Head blackish brown, each feather edged with yellowish brown; upper surface exhibiting a mixture of the pale grey feathers that mark the winter plumage with the darker, or nestling feathers; cheeks and sides of the neck pale brown mixed with grey; breast grey spotted with black; belly white with large black spots; vent and under tail-coverts white."

The Plate represents two adults, one in the winter and the other in the summer plumage, of the natural size.

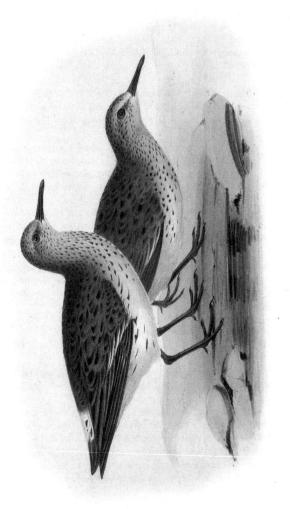

SCHINZ'S SANDPIPER.
Tringa Schinzii. *(Bonap.)*

Drawn from Nature & on Stone by J & E Gould.

Printed by C. Hullmandel.

SCHINZ'S SANDPIPER.

Tringa Schinzii, *Bonap.*

THROUGH the kindness of Sir Rowland Hill, Bart., we are enabled to add this species to the list of British Birds; a single specimen killed at Stoke Heath near Market Drayton, Shropshire, a few years since, is now in that gentleman's collection. We have compared the individual from which our figure is taken, with others killed in America, between which we could discover no difference. Its shorter bill and white rump will at all times serve to distinguish it from the other European members of the group. We believe that the continent of America is the true habitat of this species, and that its occurrence in England must be attributed to accidental causes.

M. Temminck has forwarded us specimens of the *Tringa Schinzii*, Brehm, which he informs us were received from M. Brehm himself. On examining these specimens we cannot however but express our misgivings as to their specific value, resembling as they do in every particular the Dunlin (*Tringa variabilis*) of our island: the only difference we could discover between M. Brehm's bird and examples of the Dunlin killed in England was that the former were rather smaller in size; but we doubt not that among the numerous examples of the Dunlin brought to our markets we could find males equally diminutive. The bird here represented, and which is very distinct from the Dunlin, has been considered by American naturalists as the *Tringa Schinzii* of Brehm, in consequence of which much confusion has arisen, which confusion will perhaps be removed by allowing the name of *Schinzii* to remain for the present bird, and consequently to be the *Tringa Schinzii* of Bonaparte and not of Brehm, whose bird we firmly believe to be synonymous with *Tringa variabilis*.

"They are," says Mr. Nuttall, "either seen in flocks by themselves or accompanying other Sandpipers, which they entirely resemble in their habits and food, frequenting marshy shores, and borders of lakes and brackish waters. They associate in the breeding-season, and are then by no means shy; but during autumn, accompanying different birds, they become wild and restless. Their voice resembles that of the Dunlin, but is more feeble; and they nest near their usual haunts." The eggs are four in number, smaller than those of the Dunlin, and of a yellowish grey spotted with olive or chestnut brown.

Crown of the head, neck, wings, and tail dark brown with paler margins; centre of the back and scapularies blackish brown tipped with pale brown and margined on each side with rufous; primaries dark brown, with white shafts; rump pure white; throat and all the under surface whitish; the lower part of the neck, breast, and flanks ornamented with numerous oblong spots of dark brown; bill and feet black.

Our figure is of the natural size.

BROAD-BILLED TRINGA.
Tringa platyrhyncha. (Temm.)

Drawn from Nature & on Stone by J. & E. Gould.

Printed by C. Hullmandel.

BROAD-BILLED TRINGA.

Tringa platyrhyncha, *Temm.*

Le Bécasseau platyrhinque.

THE specific name of *platyrhyncha* is given to this species in order to indicate the breadth and flatness of the beak. Like many others of its genus, it has been confounded with various species, and it is but recently that it has been extricated from the confusion in which it was involved.

The high northern regions appear to be its true habitat, whence it passes annually southwards along the rivers of the eastern portion of the Continent; it is said also to be common on the borders of the lakes of Switzerland, particularly in spring. It is a species of some rarity, and its history is but imperfectly understood. Like most of the *Tringæ*, it is subject to a change of plumage in summer and winter. Our Plate represents it in its autumnal livery, when the red markings of the upper surface have disappeared, in consequence of the edges of the feathers wearing away, while at the same time the rufous tint which covers the eye-streak and the face has given place to a dirty white.

Of its habits, manners, and nidification we have been unable to gain any information.

The sexes are alike in the colouring of their plumage: that of the individual we have copied, and which we believe to be an adult in its autumnal plumage, is as follows:

Sides of the face and neck white dotted with brown; an obscure stripe of brown from the base of the beak to the eye; top of the head and whole of the upper surface black, each feather being more or less edged with tawny white; throat and chest white marked with red and brown; abdomen white; primaries dark brown; the outer tail-feathers greyish brown; feet dark olive brown.

The figure is of the natural size.

LITTLE SANDPIPER.

Tringa minuta; (Leisler)

Drawn from Life & on Stone by J.& E. Gould.

Printed by C. Hullmandel.

LITTLE SANDPIPER.

Tringa minuta, *Leisler.*

Le Becasseau echasses.

This beautiful species, which we have figured in various states of plumage, is, with one exception only, the smallest of the British Sandpipers, and although it is nowhere numerous, is yet occasionally obtained on various parts of our coast, where it feeds and flies in company with small flocks of the Purre and the Sanderling, but sometimes occurring alone.

It is not uncommon about the margins of large fresh-water lakes on the continent of Europe generally; and specimens received from India have proved, on comparison, to be identical with those of Germany, France and Holland.

As this little bird constantly assumes at the breeding season a plumage peculiar to that period, distinct in colour from that which it bears during winter, and has besides been confounded by several authors with one European, and one extra-European species, we shall endeavour to supply descriptions of *Tringa minuta* under its various appearances, which, with the figures, will enable our readers to distinguish it at any season.

The whole length of the bird is about six inches, females being larger than males; beak straight, rather shorter than the head, black; all the upper parts of the body ash-colour, with a dusky brown streak in the line of the shaft of each feather; sides of the chest ash-colour, tinged with reddish brown; a brown line from the eye to the beak; front of the neck, throat, middle of the breast and all the under parts pure white; quill-primaries dusky black; lateral tail-feathers ash brown, the two middle ones darker brown, these last and the outer ones on each side longer than the others, giving to the tail the appearance of being doubly forked; legs and feet black, *tarsi* measuring ten lines in length, bare portion above the *tarsi* short. This is the appearance of the plumage in winter.

In the breeding season the top of the head is spotted with black and bright red; cheeks, sides of the neck and breast rufous, marked with small angular brown spots; under parts white as in winter; feathers of the back, the scapulars, wing-coverts, tertials, rump and two middle tail-feathers deep black in the centre, but nearly all have a broad border of bright red; some few feathers about the wings retain through the summer the ash colour peculiar to winter, remaining unchanged; lateral tail-feathers ash brown; legs and feet black.

Young birds of the year have the feathers on the top of the head spotted with black, and edged with reddish buff; those on the upper surface of the body and wings margined with pale buff-coloured white; the feathers on the sides of the neck, the scapulars and tertials bordered with yellowish red; in other respects like the parent birds.

We have figured these small Sandpipers of their natural size; the representation on the right side of our Plate is in the plumage of winter; that in the middle is in the perfect plumage of summer; the figure on the left is that of a young bird of the year.

Their food consists of small worms, aquatic insects, and minute crustacea.

But little is known of the nidification of this interesting little Tringa, but we have it in our power to add a description of its egg. This in its colour and markings is very like the egg of *Tringa hypoleucos*, but much smaller, measuring one inch one line in length and nine lines in breadth, the ground colour reddish white, spotted and specked with dark red brown.

TEMMINCK'S TRINGA.
Tringa Temminckii. (Leisl.)

Drawn from Nature & on stone by J & E Gould.

Printed by C Hullmandel.

TEMMINCK'S TRINGA.

Tringa Temminckii, *Leisl.*

Le Bécasseau Temmia.

THE *Tringa Temminckii* is the least of its tribe yet discovered : it is a species possessing many synonyms, and has been frequently confounded with its near ally the *Tringa minuta,* from which it differs in being more diminutive in size ; in never, as far as we have observed, obtaining the red colouring of the upper surface ; and in possessing much shorter tarsi, which are always olive green instead of black.   It also differs consi-derably from *T. minuta* in its habits, giving preference to inland creeks and muddy shores, rather than to the open shingly beach, which is known to be the favourite resort of the latter.   Although tolerably common on our coast during spring and autumn, we have no authentic account of its ever breeding with us ; yet from the circumstance of numbers of immature birds having come under our notice, there can be no doubt that this delicate species, as well as the Dunlin and many others of its race, rear their young in some of our more secluded and extensive marshes.   On the Continent its range appears to be universal, that is, on every line of coast favourable to its habits, and wide rivers and lakes offering a congenial habitat.   Europe, however, is not the extent of its range, as is proved by its being dispersed over the North of Africa and the greater portion of Asia.   Specimens have been received in collections from the high range of the Himalaya.

The summer plumage of this species is much darker than that of winter, the back and whole upper surface being covered with numerous markings of blackish brown.   The young of the year may at all times be distinguished by the semilunar edgings of black and grey which border the tips of each feather.

The food consists of flies and other insects, worms, and molluscous animals.

The female is rather larger than the male, but in their colouring the sexes offer no difference.

Forehead, top of the head, and whole of the upper surface greyish brown, the centre of each feather being blackish brown ; over the eye an indistinct line of white ; from the bill to the eye a pale brown streak ; chin and throat white ; sides of the neck and the breast greyish brown, with numerous small spots of dark brown ; primaries dull brown slightly edged with white ; secondaries and greater wing-coverts dull brown very slightly tipped with white ; shaft of the first quill white, the others dull brown; belly, vent, and under tail-coverts white ; tail consisting of twelve feathers, of which the six central ones are greyish brown, and the three outer ones on each side white ; bill blackish brown ; legs light olive brown ; claws black.

We have figured an adult and young bird of the natural size.

PURPLE SANDPIPER.
Tringa maritima. (Brünn.)

Drawn from Nature & on Stone by J & E Gould.

Printed by C. Hullmandel.

PURPLE SANDPIPER.

Tringa maritima, *Brunn.*

Le Becasseau violet.

" The locality of this species," says Mr. Selby, " being strictly confined to the rocky coasts of the ocean, and seldom found upon the flat and sandy shores, (the usual resort of most of the maritime scolopaceous birds,) has occasioned its falling less frequently under the notice of ornithologists, and its history has been consequently involved in much obscurity, and there is some difficulty in collating the synonyms under which it has been described by different authors."

The most remarkable feature presented by this species of Sandpiper consists in the great difference of colour between it and the rest of the genus, the plumage during a great portion of the year, and especially the breeding-season, having a rich violet lustre ; we have also seen specimens exhibiting traces of the barred markings of black and red, so conspicuous in the Knot (*Tringa Canutus*), to which species the Purple Sandpiper evidently bears a close affinity. The specimens referred to as resembling the Knot were, we must observe, from the region of the Arctic circle, whither this bird is supposed to retire, for the purpose of incubation, when it leaves us in April, and from whence it again returns to the temperate portions of Europe early in autumn, appearing in our island in October, and frequenting the rocky shores, particularly promontories, artificial jetties, and embankments. On the Northumberland coast and in the Fern Islands, Mr. Selby informs us it is very common, and he further remarks that he has met with the young in the month of June, a circumstance which proves that at least occasionally it breeds in our island.

Like many other species of the genus it congregates in small flocks, and has the same wheeling flight which distinguishes the Dunlin, &c.

Its food consists of small shelled mollusca, marine plants, and minute crustacea.

The Purple Sandpiper appears to be very widely distributed, at least over the northern portions of the globe, being common in the northern parts of America, as well as those of Europe and Asia.

In winter the head and neck are greyish black tinged with brown ; orbits, eye-streak, and chin greyish white ; breast grey inclining to brown, many of the feathers being darker in the centre and margined with white ; belly and under tail-coverts white, streaked and spotted with dark brown ; back and scapulars greyish black with purple reflections, and each feather margined with grey ; wing-coverts greyish black margined and tipped with white, forming a bar across the wings ; secondaries nearest the tertials almost wholly white, the rest only tipped with white ; rump and upper tail-coverts blackish brown ; middle tail-feathers greyish black ; outer ones lighter grey margined with white ; bill reddish orange at the base ; blackish at the tip ; legs and feet ochreous yellow.

In summer the whole of the plumage becomes darker, the purple hue more conspicuous ; the feathers on the head are margined with greyish white, and the spots on the breast are more distinct.

In the young the whole of the plumage is of a dull greyish black, margined with dirty yellowish brown ; the sides of the neck and breast are grey, with darker streaks ; and the flanks and under tail-coverts are streaked longitudinally with deep ash grey.

We have figured an adult of the natural size.

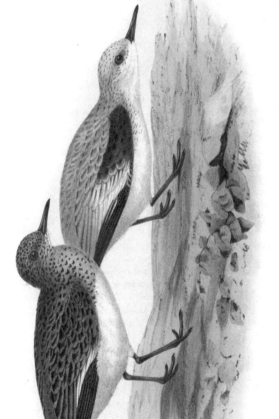

SANDERLING.

Arenaria Calidris. /Meyer/.

Drawn from Nature & on stone by J.&E.Gould.

Printed by C.Hullmandel.

Genus ARENARIA, *Bechst.*

GEN. CHAR. *Bill* as long as the head, straight, slender, semi-flexible, compressed at the base, with the tip dilated and smooth. *Nostrils* lateral, basal, narrow, longitudinally cleft in the nasal furrow, which extends to the dertrum or nail of the bill. *Wings* of mean length, acuminate, with the first quill-feather the longest. *Legs* of mean length, naked above the tarsal joint. *Feet* three-toed, all the toes directed forwards, with a very small connecting membrane at their base.

SANDERLING.

Arenaria Calidris, *Meyer.*

Le Sanderling variable.

THE Sanderling is one of the few birds whose distribution may be considered as almost universal, a circumstance probably to be accounted for by the great powers of flight and the extreme high northern latitudes to which it is known to retire for the purpose of breeding, from whence it would appear to radiate not only over the sea-shores of the Old World, but also over those of the New, extending its migration during winter to within the tropics. It is very plentiful in Brazil, whence we have received specimens which do not present the slightest difference either from those of our own island or from those of Africa and Asia.

From the dreary regions of a northern clime they commence an early return to more temperate latitudes, and it not unfrequently happens that the Sanderling may be observed on our shores as early as the month of August, " at which time," says Mr. Selby, " I have killed several individuals upon the Northumbrian strand. These have generally been the young of the year, and are probably stragglers from flocks of the earliest broods which were destined to reach more southern latitudes, as the great body that resorts to our shores and those of the opposite continent, seldom arrives before the middle of September."

The actions of the Sanderling while on the ground are characterized by the greatest activity, and bear a striking resemblance to those of many of the smaller *Charadriadæ*, among which, particularly the Ring Dottrel, it is often associated.

Its food consists of insects of various kinds, but more particularly the smaller *Coleoptera*, in the capture of which it displays the greatest agility. Its flight is rapid and vigorous ; hence it is enabled to perform considerable migrations in a comparatively short space of time.

There is no member of the family to which the Sanderling belongs that exhibits a greater change of plumage than takes place in this species ; in fact, were we not acquainted with the peculiar law relative to the members of this group, and had two examples of this bird been presented to us for the first time, one in the plumage of summer and the other in that of winter, we should undoubtedly have pronounced them to have been distinct species.

The nest and eggs are at present unknown.

In winter the forehead, sides of the neck, and all the under surface are pure white ; crown, nape of the neck, back and scapulars ash grey, with the shafts blackish brown ; secondaries brown, with white tips which form a bar across the wings ; outer webs of the greater quills deep brown, and their shafts white ; middle tail-feathers brown margined with white ; outer ones greyish white ; bill and legs black.

In summer the crown and the forehead are black, each feather margined with pale reddish brown and white ; throat, neck, and breast mingled reddish brown, ash grey, and brownish black ; back and scapulars reddish brown irregularly blotched and spotted with black ; greater coverts blackish brown margined and tipped with white, forming a bar across the wings.

The young before the first moult have the forehead, a streak over the eye, cheeks, and throat white ; at the lower part of the neck a band of yellowish white passing into ash grey ; all the under surface white ; crown of the head black margined and spotted with pale buff ; nape pale grey with streaks of a darker tint ; back and scapulars black margined and spotted with white ; tertials brown margined with greyish white ; legs blackish grey.

We have figured two birds, one in the summer, the other in the winter plumage, of the natural size.

RED-NECKED PHALAROPE.
Phalaropus Hyperboreus. *(Leith.)*

Drawn from Life & on Stone by J & E Gould.

Printed by C Hullmandel.

Genus PHALAROPUS, *Briss.*

GEN. CHAR. *Beak* straight, depressed at its base; both *mandibles* furrowed as far as the point, the tip of the upper one obtuse, and bending over the under, which is pointed. *Nostrils* basal, lateral, oval, encircled by a membrane. *Feet* moderate, slender. *Tarsi* compressed. *Toes*, three before and one behind, those in front united as far as the first joint by a membrane which is continued to their extremities in indented festoons; the hind toe having only the rudiments of a membrane. *Wings* lengthened; the first *quill-feather* longest.

RED-NECKED PHALAROPE.

Phalaropus hyperboreus, *Lath.*

Le Phalarope hyperboré.

THE two species of Phalaropes which are recognised as natives of Europe, have in their different stages of plumage received various synonyms; and the present species, from its more lengthened and attenuated bill, has been separated by M. Cuvier, and advanced to rank as a genus, which he has designated *Lobipes*. With the views of this great naturalist we do not in the present instance concur; being unwilling that birds agreeing so closely as these in habits, manners, and food, should be separated: though it must at the same time be confessed that as the modification the bird here figured exhibits in the structure of the beak, points out a degree of affinity to the genus *Totanus*, so on the contrary the other species evinces an approach to the *Tringas*, *T. hypoleucos* for example; a bird which although it does not swim, except from necessity, is certainly endowed with that power beyond its congeners. Retaining, however, the two European species under one and the same genus, we may proceed to observe, that the Red-necked Phalarope is the least of the two, and that the elegance of its form, together with the grace and ease of its actions on the water, cannot but excite the admiration of every lover of Nature. It more particularly inhabits the northern portion of the globe, being found both in Europe and America, frequenting the shores of the sea and large sheets of water whether fresh or salt, but more especially the latter. In the British dominions,—Scotland, and its northern and western Isles, are the most frequent places of resort, where it also breeds; the specimens from which our figures were taken having been collected among the Shetland Islands, in their mature and breeding plumage during the season of 1832, by Mr. Dunn of Hull, who informs us that they were by no means uncommon, and that their familiar and unsuspecting habits rendered them easy of acquisition. Their nests, which have been found among the lochs of Sanda, as described by Mr. Salmon, were placed in small tufts of grass growing close to the edge of the water; the eggs four, one inch two lines in length, and ten lines and a half in breadth, olive-brown spotted and specked with brownish black. M. Temminck states that in Germany and Holland this bird is of rare occurrence.

As we might expect from the lengthened form of the wings, the power of flight which the Red-necked Phalarope possesses is very considerable; nor is it less endowed with facilities for swimming, not only upon the smooth surface of lakes and ponds, where it is sometimes seen, but also upon the rougher billows of the ocean far from shore, where it finds itself quite secure. On land it does not display that lightness and activity which characterize the *Tringa* in general; in its lobed feet, however, it possesses an advantage over that tribe in being able to walk on the soft and oozy mud which covers the sides of creeks and estuaries, among which it finds its principal food, consisting of insects, worms, and minute mollusca.

The changes of plumage which this bird undergoes are but little understood; we know, however, that the young differ materially from the adult birds, having a lighter colouring of plumage, wanting the red on the sides of the neck, and all their feathers being margined with greyish white. M. Temminck, as well as other naturalists, has fallen into an error respecting the sexual differences which characterize the Phalaropes, the Sandpipers and Plovers, the Ruff excepted,—viz. in considering the largest and richest-coloured birds to be males, whereas the contrary is in reality the case. This law appears to prevail with most of those birds that produce but one brood of young during the summer, and the females are further remarkable for laying very large eggs in proportion to the size of the bird.

We take our description from adult specimens now before us. The whole of the head, the back of the neck, the breast and flank, are of a dark ash colour; throat, belly and vent, white; between the breast and throat intervenes a broad patch of beautiful chestnut-red; the remainder of the upper plumage of a brownish black, the feathers having a rufous margin; secondaries tipped with white, which forms a band across the wings; bill black; irides brown; feet olive-brown. Length about six inches.

We have figured both sexes in their summer plumage; the female will be readily distinguished by her more brilliant colour and larger size.

GREY PHALAROPE.
Phalaropus platyrhynchus, (Temm.)

Drawn from life and on Stone by J. & E. Gould.

Printed by Hullmandel.

GREY PHALAROPE.

Phalaropus platyrhynchus, *Temm.*

Le Phalarope platyrhynque.

THIS bird can no longer be ranked among the rarities of our British catalogue, from its frequent occurrence and the many instances of its capture of late years in this country, particularly in its winter plumage, in which state it is known by the name of Grey Phalarope ;—a mode of nomenclature which, referring to one state only, we consider by no means appropriate to birds that undergo various periodical changes of plumage. Indeed, were it not for adding to the list of synonyms, already too numerous, we should have ventured to designate the present species " the Broad-beaked Phalarope," in order to distinguish it from its congeners, from which it differs so much in this single character ; we retain, however, the name of Grey Phalarope, as it is generally used by British Ornithologists.

The native habitat of this species is well known to be the regions extending within the limits of the Arctic circle, where it takes up its summer abode, migrating as the severity of winter comes on to more temperate climes, and dispersing singly or in pairs throughout most of the countries of Europe, especially the British Islands ; nor is it less abundant in many parts of Asia, as well as of America, from the northern towards the intertropical regions. Although the powers of wing, which enable the Phalaropes to make extensive migrations, are very great, still we do not look for their periodical visits with that degree of certainty and regularity which characterizes the migrations of birds in general. The places, moreover, which it not uncommonly chooses for residence during its sojourn with us, are such as would possess for it, according to our ideas, but little attraction ; thus, for instance, it will often continue for weeks together, if unmolested, about a farm-yard pond or mere puddle, manifesting a familiar and unsuspecting disposition, and allowing itself to be approached with freedom ; it does not, however, appear to confine itself much longer to one spot, but after remaining at a certain place from one to three or four weeks, suddenly departs, if on the approach of spring, towards the north, and in autumn towards the south ;—every European country in fact appears to be equally visited, although at uncertain and often long intervals.

Like the other species, it is an admirable swimmer, taking its food on the surface of the water with the utmost agility and address ; indeed it appears to seek its nourishment there alone, and may be watched, while assiduously engaged in this occupation, displaying a thousand graceful attitudes and manœuvres.

Though usually seen in England in its grey or winter livery, it sometimes occurs in an intermediate state during the progress of change, and occasionally, though very rarely, in its red or summer plumage. Our most frequent visitors of this species are young birds of the year, which make their appearance during autumn, and are then for the first time putting forth the delicate grey feathers of the back, which they carry through the winter.

We transcribe from the published account of these birds by Captain Edward Sabine, some particulars of the difference in the size of the sexes, and of their plumage, which had been previously unnoticed. "Average length of males 7 inches 6 lines ; extent 16 inches 2 lines ; weight 1¾ ounce. Females 8 inches 4 lines ; extent 17 inches ; weight 2 ounces. The breeding plumage of the male corresponds minutely with the description which Temminck has assigned to both sexes : the female has the forehead, crown and hind head a uniform deep sooty black, without intermixture of orange or red : the band which passes through the eye is a pure white, and is larger and better defined than in the male, including more space above and in front of the eye ; the black predominates in the back and scapulars, the orange bordering of the feathers being smaller and much lighter than in the male ; the under plumage is of a deeper and richer brick-red colour, and is unmixed with white feathers for a much longer portion of the season ; the female bird attains her perfect plumage earlier in the year, and retains it longer than the male, which is also the case with several other of the Northern birds."

Few birds differ so much in the winter and summer states of their plumage as the present, the transition being from a strong reddish brown to a delicate silver grey.

The nest is unknown, or, at least as far as we are acquainted, is undescribed ; the eggs are one inch two lines and a half long, and eleven lines across, of a greenish stone colour, spotted and specked with black.

COOT.

Fulica ater. (*Linn.*)

Genus FULICA.

GEN. CHAR. *Bill* shorter than the head, strong, straight, subconical, compressed, much higher than broad, the culmen of the upper mandible distended into a broad shell-like plate, which extends over a portion of the forehead. *Nostrils* concave, pierced in the membrane of the mandibular furrow near the middle of the bill, pervious, linear, and oblong. *Wings* with the second and third quill-feathers the longest. *Tail* short. *Legs* of mean length and strength; naked above the tarsal joint. *Feet* four-toed, three before and one behind. *Toes* long, united at the base, and lobated, the inner one with two, the outer one with four, distinct round membranes: middle toe longer than the tarsus.

COOT.

Fulica ater, *Linn.*

Le Folque macroule.

THE Coot is indigenous to our islands, residing on all large sheets of water, whether flowing or stagnant, but giving preference to those overgrown with rushes and margined with a belt of thick reeds and luxuriant vegetation. It abounds in equal numbers throughout the continent of Europe, particularly in Holland, France, and Germany. In the secluded situations above mentioned it prepares early in the spring for the work of incubation, building a large, strong, and solid nest composed of rushes, various grasses, and aquatic plants. The nest thus put together rises above the level of the water, the mass of compacted materials in some cases resting on the bottom, where the shallowness of the water will admit, but is more frequently intermingled with the tufts of vegetation which grow in abundance on the water's edge and partially conceal it from view. On this raft the female deposits her eggs, which are of a brownish white spotted with dark brown, and from seven to ten in number, and there patiently performs her allotted task. The young when first excluded are clothed with a black down, and actively take to the water, attended by their assiduous parent, who may be often seen thus leading her tribe of nestlings in the earnest search for food, which consists of seeds, aquatic plants, insects, and mollusca.

When winter covers the ponds, lakes, and canals with ice, thus cutting off every needful supply, the Coot leaves its secluded quiet haunts of summer, and seeks the wide stream of the larger rivers, venturing even as far as their embouchures in the sea. At Southampton, multitudes annually visit the river during this season, disappearing on the approach of spring; and it is generally observed, that from October the places where they have taken up their summer abode are deserted till the month of April, when they again make their appearance.

It seems almost needless to say that few birds swim more easily or gracefully than the Coot; it also dives with considerable facility: on wing, however, it is slow and embarrassed, and, indeed, seldom rises unless so pressed that no other means of escape present themselves. On land it trips along with great facility, and, indeed, may be often observed reposing on the bank, or, like the Gallinule, traversing up and down in quest of worms and slugs, which it devours with much avidity. If surprised, it immediately plunges into the water and makes its way as rapidly as possible to the dense covert of reeds or rushes, where it is effectually concealed.

No external difference characterizes the sexes; nor do the young of the year exhibit any difference, except that the frontal plate is imperfectly developed.

The general plumage is deep greyish black, with a tinge of blue on the under surface; bill and frontal plate white; irides scarlet; naked part of the tibiæ orange; tarsi and toes olive green, the former tinged with yellow.

Our Plate represents an adult bird rather less than the natural size.

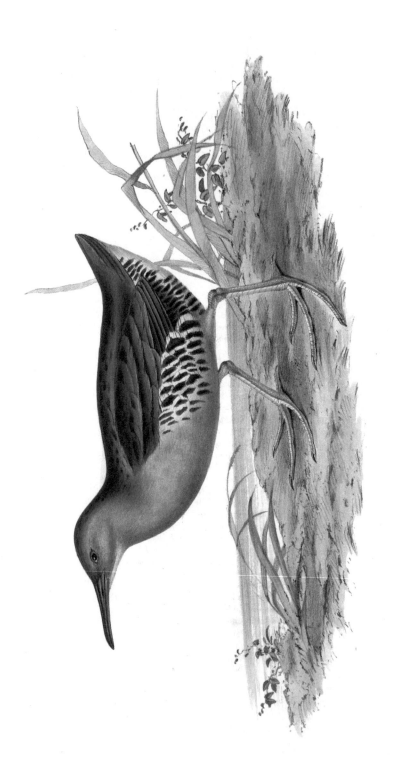

WATER RAIL.

Rallus aquaticus. (Linn.)

Drawn from Life & on Stone by J & E Gould.

Printed by C. Hullmandel.

Genus RALLUS, *Linn.*

GEN. CHAR. *Beak* slender, longer than the head, slightly arched or straight, compressed at its base, cylindrical at its point; the upper *mandible* furrowed. *Nostrils* communicating, lateral, opened longitudinally in the furrow, partly closed by a membrane. *Tarsi* long and strong, naked above the knees. *Toes* three before and one behind; the anterior ones divided; the posterior articulated upon the tarsus. *Wings* moderate and rounded; the third and fourth *quill-feathers* longest.

WATER RAIL.

Rallus aquaticus, *Linn.*

Le Rale d'Eau.

THE Water Rail is very generally dispersed over Europe, but abounds principally in the low flat lands of Holland, France and Germany, where fresh-water morasses, swamps, and rivers afford a congenial and native habitat; and although never observed in any abundance in the British Islands, the apparent scarcity must be attributed rather to its cunning and retired habits than to its being really a rare bird. Except when closely pressed, the Water Rail seldom takes to flight, but evades pursuit by quietly yet quickly traversing the bottoms of thick-set reed-beds and banks overgrown with luxuriant vegetation, bordering the sides of pools and ditches, where it finds a covert, through which its slender and compressed form enables it to pass with the greatest facility; besides which it possesses the power of swimming and diving, both of which materially aid its escape. Without denying the possibility of this bird being migratory, we have the strongest reason to believe that numbers remain with us during the whole of the year, frequenting during the summer season fen land, morasses, ponds, and ditches, about which it incubates; resorting on the approach of winter to the sides of our large streams and rivers. Its nest is composed of rushes and vegetable fibres, closely concealed among herbage, at a little elevation from the water; its nidification, in fact, closely resembles that of the Moorhen. Its eggs are of a yellowish white colour, marked with spots of red brown. Its food consists of worms, snails, soft insects and their larvæ, which abound in swampy places; vegetable substances also form a part. The young when first excluded from the egg are covered with black down, and are observed to be in perfect possession of the powers of swimming, and providing for their own safety and subsistence; remaining, however, under the parent's care and protection. In a short time their plumage undergoes a change; the feathers characterizing the species advance through the down, and they then nearly resemble the adult bird, but are to be distinguished by the breast and under parts being of a reddish brown, and the markings of the flanks more obscure and undefined. The sexes are alike in plumage, but the male is generally the largest. The throat is whitish; the sides of the head, neck, breast and belly of a blueish ash; the upper surface brown, the centre of each feather black; the feathers of the flanks are barred transversely with clear black and white; the under tail-coverts white; the beak red at its base, becoming gradually black towards the tip; irides reddish orange; feet and toes light brown.

We have figured an adult bird of its natural size.

HYACINTHINE PORPHYRIO.
Porphyrio hyacinthinus. *(Temm.)*

Genus PORPHYRIO.

GEN. CHAR. *Bill* shorter than the head, strong, hard, thick, concave, nearly as high as long, culmen of the superior mandible depressed and dilated upon the forehead. *Nostrils* lateral, placed near the ridge, and pierced through the mandible. *Feet* strong. *Toes* long; each furnished with a fine lateral membrane. *Wings* moderate, the third or fourth quill-feather the longest.

HYACINTHINE PORPHYRIO.

Porphyrio hyacinthinus, *Temm.*

Le Talève Porphyrion.

THE birds forming the restricted genus *Porphyrio* may be readily distinguished from the Gallinules (*Gallinula*) by the greater depth and richness of the colour of their plumage, by the extraordinary development of the feet, and by the robust form of the bill. Although the number of species is somewhat limited, they are widely distributed over the tropical portions of the Old World,

Independently of the southern and eastern parts of Europe, the marshes of which are the places of constant resort for this beautiful bird, its range is extended over a great portion of Africa to the south, and as far as the mountains of the Himalaya to the east. In Europe it is especially abundant in the Grecian Archipelago, the Levant, and the Ionian Islands; it is less common in Dalmatia and Sardinia. The southern provinces of Hungary and Russia, and the borders of the Caspian Sea, may also be enumerated among its European localities.

Like the Water-hen, or Common Gallinule, it dwells on the borders of rivers and in all marshy situations. In its food it is partly herbivorous, feeding on various kinds of marine vegetables; still, as the robust and hard character of its bill implies, it is destined to live upon other food, and hence we find it frequently giving a preference to hard seeds and grain, to which are added snails, frogs, and other aquatic animals.

Although its form would seem to deny the fact, its actions and appearance on the land are both elegant and graceful. It is extremely quick in all its movements, running with ease and swiftness; and from the great expansion of its feet it is enabled to pass with facility over soft oozy mud, aquatic herbage, &c.: but although much agility characterizes this species on land, its aërial evolutions are heavy, and apparently performed with considerable difficulty.

The sexes offer no difference in the colour of their plumage. They breed in marshes, much in the manner of the Common Gallinule, giving preference to the sedgy parts of the morass and partly inundated rice-fields, where it constructs a nest of aquatic plants, and lays three or four white eggs that are nearly round.

Bill fine red; legs and feet fleshy red; irides lake red; cheeks, throat, sides of the neck, and chest turquoise blue; remainder of the plumage deep dull indigo blue, having the edges of the greater and lesser coverts of the wings lighter in colour and more brilliant; under tail-coverts white.

The Plate represents an adult of the natural size.

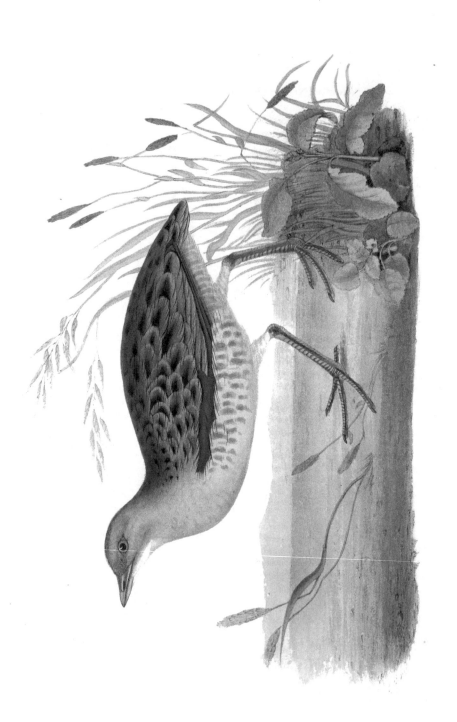

LAND RAIL.

Gallinula crex. *(Lath.)*

Genus GALLINULA.

GEN. CHAR. *Bill* shorter than the head, compressed, conical, higher than broad at the base, the ridge advancing on the forehead, and dilating itself in some species into a naked plate; *mandibles* of equal length, the points of both compressed, the upper slightly curved; the *nasal grooves* very large, communicating. *Nostrils* lateral, longitudinal in the middle of beak, partly closed by membrane. *Legs* long, naked above the knee. *Toes,* three before, and one behind. *Wings* moderate, and slightly rounded.

LAND RAIL.

Gallinula crex, *Lath.*

Le Rale de Gennet.

THE Land Rail appears to be extensively spread over the whole continent of Europe; it is very abundant in Holland, and not uncommon in France and Germany. It is a migratory species, arriving with us about the latter end of April or the beginning of May, when it scatters itself in pairs over the whole of the British Isles. Its habits are extremely shy and retiring, selecting for its places of abode grassy meadows, fields of young corn, ozier-beds, and marshy grounds, seldom allowing itself to be seen; and were it not for the peculiar note of the male, which consists of a singular grating *monotone,*—sometimes sounding as if beneath one's feet, and again appearing as if uttered at a distance,—its presence would not be betrayed. In these its favourite places of resort and concealment it carries on the process of incubation, constructing its nest on the ground, and occasionally on small hillocks; the nest being composed of slender flags or grasses; the female laying from eight to twelve eggs, rather less than those of the Moor-hen, to which in the markings they bear some resemblance, of a yellowish-white, covered with dull rust-coloured spots. The young when hatched are covered with a black down, and are soon able to follow the parent birds, attaining by the commencement of the shooting season nearly the adult size and plumage, when they are often killed by the sportsman, and much sought after by the epicure, being esteemed a great delicacy.

It is with difficulty forced to take wing, but runs with great rapidity before the pursuer, whom in general it readily eludes.

Its flight is very short and embarrassed, with the legs hanging down; and it evidently trusts for safety more to its activity on the ground, and to the seclusion afforded by the herbage beneath which it crouches.

In autumn, when the fields and meadows no longer afford cover for its concealment, it retires to brakes and thickets, and its note is seldom heard. In the latter part of October they commence their migration southwards, passing over to the Continent; leaving occasionally a few feeble or wounded birds behind, which remain with us during the winter.

The male and female are alike in plumage: their food consists of insects and their larvæ, such as grass-hoppers, &c., as also worms, snails, vegetables, seeds, and grain.

The bill and legs are flesh-coloured; irides hazel; over the eye a large ash-coloured mark, which extends towards the occiput; the top of the head and the whole of the upper surface of the body of a rufous brown, the feathers of the back having a black mark in the centre; the shoulders and quill-feathers of a light chestnut; the sides light brown, marked with darker transverse bars; breast ash-colour, inclining to a lighter tinge on the under parts.

Length nine inches and a half; weight seven or eight ounces.

The figure in our Plate is that of an adult in its spring plumage.

COMMON GALLINULE.
Gallinula Chloropus. (Lath.)

COMMON GALLINULE.

Gallinula chloropus, *Lath.*

La Poule d'Eau ordinaire.

THIS common species appears not only to be dispersed over the whole of Europe, but extends its range over the greater portion of Africa and India; and, in fact, like the Peregrine Falcon and Barn Owl, it may be said to be universally distributed over the globe: it is even questioned among some of our most able naturalists whether those from tropical America, China, and the islands of the Pacific, which exhibit the most trifling marks of difference, should not be considered as identically one and the same species. In the British Islands it dwells in rivers, ponds, sedgy districts, and all low marshy situations. During the severities of winter, when all our inland waters are frozen over, it retires to the larger streams and rivulets, which afford it during the rigorous weather not only a better protection against the sportsman, but also a supply of food, which could not be procured on the banks of its favourite pond or accustomed residence. Although its long and thin toes would appear to be but little adapted for such a purpose, it nevertheless possesses the greatest facility for diving, which power it not unfrequently makes use of for the purpose of obtaining water-snails, insects, and their larvæ, which, with tender weeds and grasses found at the bottom of the stream, constitute part of its food. In less rigorous weather it may frequently be seen on land, particularly in meadows and grass-fields, feeding upon worms and insects, and when thus observed its actions are both elegant and graceful; if unmolested it soon becomes less shy and retiring, and adds considerably to the life of the landscape. Its flight is heavy and awkward, and seems to be performed with great exertion. One circumstance respecting this familiar bird appears to have escaped the notice of most ornithologists, we allude to the fact of the female being clothed in a dark and rich plumage, and having the base of the bill and frontal shield of a bright crimson red tipped with fine yellow; her superiority in these respects has caused her to be mistaken for the male, which, contrary to the general rule, is at all times clothed in a duller plumage, and has the upper surface more olive than in the female; the bill is also less richly tinted. We were first led to notice this fact in consequence of observing the birds sitting or rising from the nest to be those whose richly coloured bills had induced us to believe them to be males, and which the dissection of a great number of individuals has now fully proved to us to be the females. Besides this difference in colouring, the sexes vary in size, the female being about one fifth less than her mate.

The nest of the Common Gallinule is neatly constructed of flags and weeds, and is placed among the rushes in the most retired parts of the brook or pond. The eggs are from five to nine in number, of a pale yellowish brown spotted all over with red. The young, which are hatched after an interval of three weeks from the time the female commences sitting, are clothed with a black down, and so strictly aquatic are they in their habits that they take to the water the moment after they are excluded from the shell, and are in immediate possession of all the faculties requisite for obtaining their subsistence, feeding on water-insects, flies, &c. At this tender age they encounter many enemies, and require the most assiduous care of their parents to protect them from the attack of rats, weasels, and the voracious pike, which commits the most destructive havoc not only among the young of this species but also those of many other kinds of water-fowl. The young during the first autumn, although equal to the adults in size, have a much lighter plumage, the whole of the throat and under surface being then greyish white and the bill and legs olive.

The male has the bill red at the base strongly tinged with olive; the whole of the upper parts olive brown; breast and under parts dark bluish grey tinged with olive; the centre of each feather on the flanks is blotched with a large oblong patch of white, which is the colour of the under tail-coverts; irides red; tarsi and toes greenish olive, the former being encircled with a red mark immediately above the tarsal joint, which is commonly called the *garter*.

The distinguishing characters of the female and young being given above, it is unnecessary to repeat them here.

The Plate represents an adult female and a young bird of the first year, of the natural size.

SPOTTED CRAKE.
Zapornia Porzana.

Drawn from Nature & on Stone by J.&E.Gould.

Printed by C.Hullmandel.

Genus ZAPORNIA, *Leach.*

GEN. CHAR. *Beak* slender, shorter than the head, acuminated, compressed, acute; the upper mandible gradually incurved. *Nostrils* linear, lateral, placed at the base of the beak. *Neck* elongated and slender. *Legs* long, slender, cleft, with three toes in front: the hinder toe elevated from the ground at its base: the tibiæ half naked.

SPOTTED CRAKE.

Zapornia porzana.

La Poule d'Eau Maronette.

ALTHOUGH the group of which the Land Rail is the type, and the members of the present genus, approximate very closely, still they differ so much in their general habits and in their style of colouring that we are inclined to admit the validity of their separation; and although the present bird was not included by Dr. Leach in the genus he established, we conceive that it strictly belongs to it, and have consequently associated it with the two other species *Zapornia pusilla* and *Zap. Baillonii.*

With regard to their economy and habits, while the Land Rail is entirely confined to meadows and fields, the Spotted Crake and its congeners, on the contrary, are strictly aquatic, so much so, indeed, as to make the waters their constant asylum; and although not web-footed, they swim with the greatest facility. The dense vegetation along the borders of marshes and pools is the situation to which they are particularly attached: they are rarely seen on the wing, and are scarcely ever flushed unless closely pursued by a dog.

The Spotted Crake is found in the North of Asia, is particularly abundant in the northern and eastern parts of Europe, and in the British Islands is a periodical visitor, arriving early in spring and departing on the approach of the severities of winter.

"Its nest," says Mr. Selby, "is built among the thick sedges and reeds of the marshes, and from the foundation of it being frequently placed in water, is composed of a large mass of decayed aquatic plants interlaced, with the hollow neatly formed, and comfortably lined. The eggs are eight or ten in number, of a yellowish grey colour, with a tinge of pink, and with round spots of umber brown of various sizes, and with other secondary colours of a lighter shade. It feeds on worms, aquatic insects, slugs, seeds, &c.; and its flesh is sweet and well flavoured. In autumn it becomes loaded with fat, a layer of nearly a quarter of an inch in thickness covering the whole surface of its body."

The sexes have no distinguishable difference in the colouring of their plumage, nor do the young of the year offer any considerable variation in their colour or markings.

Crown of the head and the whole of the upper surface deep greenish olive speckled with white, the centre of each feather very dark; wing-coverts and secondaries spotted and crossed transversely with irregular markings of greyish white and black; primaries dark olive brown, edged with greenish olive; stripe over the eye and throat grey; sides of the neck, breast, and under surface pale greenish olive, spotted and transversely barred with greyish white bounded by black; bill red at the base and yellow at the tip; legs olive yellow.

The Plate represents an adult of the natural size.

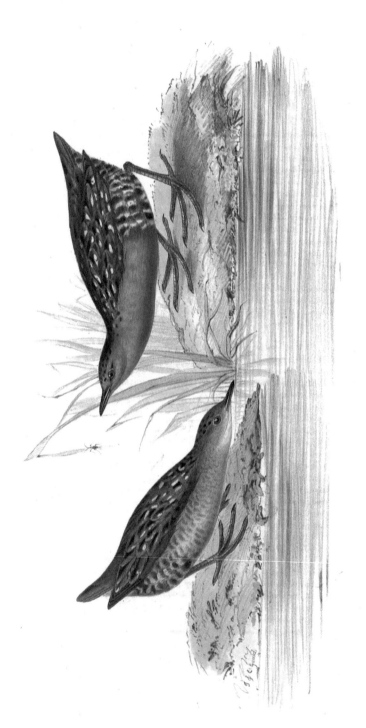

BAILLON'S CRAKE.
Zapornia Baillonii. /Leach/

Drawn from Nature & on Stone by J & E. Gould.

Printed by C. Hullmandel.

BAILLON'S CRAKE.

Zapornia Baillonii, *Leach*.

Poule-d'eau Baillon.

THE very prettily marked bird figured in our Plate is the smallest of the European Crakes, and although long known on the Continent, it is only since the days of Montagu that this species has been added to our British catalogue; it being considered as established—from the opportunities of examining both the specimens that formerly belonged to our indefatigable English ornithologist, and which are now in the British Museum, as well as the specimen which belonged to Mr. Plasted of Chelsea, also referred to in the Supplement to Montagu's Ornithological Dictionary,—that neither of those birds belong to the species now under consideration, which latter appears to have been obtained in this country more frequently than *Z. pusilla*, and of which we have examined a variety of specimens, continental as well as British, in different states of plumage. We may here mention as a particular mark of distinction, to which we have as yet seen no exception, that, when adult, *pusilla* exhibits on the upper surface but a few indistinct white marks, and those confined to a small space on the centre of the back only; in *Baillonii*, on the contrary, these white marks are very numerous, occupying several distinct situations, namely, the central line of the back, and the scapulars, wing-coverts, and tertial feathers on both sides. These white marks, placed on a black ground, forming the centre of each feather, are so conspicuous and brilliant as to have led M. Temminck originally to select the term *stellaris* for this species; but he subsequently proposed to change *stellaris* for *Baillonii*, as a compliment due to the celebrated naturalist of Abbeville; and this latter name has been received and adopted.

Baillon's Crake inhabits the southern and south-eastern portions of Europe, is rather common in Italy, and found in several provinces of France. In England it can be considered but as an occasional visitor, and has only been obtained in the south-eastern part. The most usual places of resort with this species are the banks of rivers, large lakes, ponds and marshes, where aquatic vegetation is' in luxuriance. In such situations its timid nature finds a secure retreat, and passing its small and delicate form through the thick herbage with great ease, it can rarely be made to take wing. It is said to be able to swim and dive very readily, and makes its nest near the water's edge, in which it deposits seven or eight eggs, not unlike a large olive in form, size, and ground-colour, but spotted with darker greenish brown. Its food consists of worms, slugs and insects, with portions of vegetables and seeds. A specimen in the collection of the Rev. Dr. Thackeray, Provost of King's College, Cambridge, was taken during a cold and frosty January on some ice near Melbourn, about nine miles south of Cambridge. To this spot, originally fen land, the poor bird had resorted in an inclement season to obtain a meal, but having wandered far from its native and more congenial latitude, was so exhausted by want of food or the low temperature of the season, or the combined effects of both, as to allow itself to be taken alive by the hand.

In adult males the forehead, eyebrows, sides of the neck, and the whole of the under surface are of a dark-blue grey, almost approaching to black on the belly and flanks, which are barred with white; the top of the head, back of the neck, and all the feathers on the upper surface of the body, of a rich olive brown, the centre of each feather more or less pervaded with black; those of the middle of the back, scapulars, wing-coverts, and tertials splashed with pure white; the primaries are dark brown, and extend only to the middle of the tail-feathers, which are also dark like the primaries, but edged with olive brown, the marginal markings becoming narrower as the bird increases in age. Beak green; irides hazel; legs flesh colour, but darkest in adult birds. Several examples averaged seven inches in length, from the point of the beak to the end of the tail.

Adult females differ but little from the males, except that their general colours are less vivid. Young birds have the chin and throat white, the neck, breast and belly mottled with reddish brown, dusky black and dull white; abdomen and flanks less distinctly marked by black and white, the alternate bars not so well defined, and the two colours much less decided.

We have figured an adult and a young bird of the natural size.

LITTLE CRAKE.
Zapornia pusilla. (Steph.)

Drawn from Nature & on Stone by J & E. Gould.

Printed by C. Hullmandel.

LITTLE CRAKE.

Zapornia pusilla, *Steph.*

Poule d'eau Poussin.

WE have preserved the term *Little*, given to this species by Montagu, although the bird is somewhat larger than that named after M. Baillon, which has also occurred in this country even more frequently than the subject of the present article. Indeed, except the examples obtained by Montagu, that which belonged to Mr. Plasted of Chelsea—also noticed by Montagu in the Supplement to his Ornithological Dictionary,—and a specimen taken alive in a drain in Ardwick meadows, near Manchester, in the autumn of 1807, by Mr. James Hall, as recorded in Loudon's Magazine of Natural History, vol. 2. p. 275, we are not aware of any other instances of this species having been obtained in this country. This Little Crake is, however, by no means so rare on the European Continent: it is even common in the eastern countries of Europe, in Italy and in Germany; more rare in the northern parts of France, and only occasionally taken in Holland. It principally frequents marshes, but is sometimes seen on the higher and more cultivated soils. The habits of the smaller species of Gallinules, says Montagu, are their principal security: they are not only capable of diving and concealing their bodies under water, with only the beak above the surface to secure respiration, but run with celerity, and conceal themselves amongst the rushes and flags of swampy places, and are with difficulty roused even with the assistance of dogs, depending more on concealment in thick cover than upon their wings to avoid danger. Insects, slugs, the softer aquatic vegetables and seeds are the principal food of this species. It constructs a nest among reeds, upon the broken stems of rushes and water plants, and lays seven or eight oval-shaped eggs, of a yellowish brown colour, spotted with elongated marks of darker olive brown. In the adult male, the eyebrows, cheeks, front and sides of the neck, breast and belly are of uniform slate grey, without any spots; abdomen and flanks mixed and barred with brown and white; the top of the head and all the upper parts generally olive brown, the feathers on the middle of the back much darker in colour, almost black, and varied with a few white marks, but without any white on the wings or wing-coverts; the tertials dark in the centre, olive brown at the edge; the primaries uniform dusky; under tail-coverts dark lead colour, almost black, but barred with white; beak olive green; the base orange yellow; irides reddish hazel; legs and toes olive green; length seven inches and a half.

Notwithstanding some slight differences, we believe the Olivaceous Gallinule of Montagu to be identical with the adult male bird here described; and we also consider his little Gallinule to be a young bird of this same species, which may be thus described; eyebrows and sides of the head light ash colour; throat whitish; chest and belly brownish buff, thighs and flanks ash coloured, barred with brown and white; under tail-coverts tipped with white; upper parts reddish brown; the dark space on the middle of the back varied with a few white spots; wing-coverts olive; beak olive without the orange base; eyes dark hazel; legs and feet olive. Whole length seven inches and a half. Young birds are still lighter in their general colouring; the whole of the throat and neck is whitish; the white marks on the back are very few in number, or scarcely perceptible; and the feathers on the flanks are brown, with white bars, without the ash colour.

We have figured an adult male and young bird of the natural size.

THE

BIRDS OF EUROPE.

BY

JOHN GOULD, F.L.S., &c.

IN FIVE VOLUMES.

VOL. V.

NATATORES.

LONDON:

PRINTED BY RICHARD AND JOHN E. TAYLOR, RED LION COURT, FLEET STREET.

PUBLISHED BY THE AUTHOR, 20 BROAD STREET, GOLDEN SQUARE.

1837.

LIST OF PLATES.

VOLUME V.

NOTE.—As the arrangement of the Plates during the course of publication was found to be impracticable, the Numbers here given will refer to the Plates when arranged, and the work may be quoted by them.

NATATORES.

* Named erroneously Carbo pygmæus.
† Named erroneously on the Plate Larus atricilla.

SNOW GOOSE.
Anser hyperboreus. *(Pall.)*

Printed by C. Hullmandel

Genus ANSER, *Briss.*

GEN. CHAR. *Bill* as long as the head, straight, conical, thick, higher than broad at the base, depressed and flattened towards the tip; entirely covered with a cere, except the nails at the tip, which are horny, orbiculate and convex; lower mandible narrower than the upper; the edges of both laminato-dentated. *Nostrils* lateral, placed near the base of the bill. *Wings* long, ample, tuberculated. *Legs* placed a little behind the equilibrium of the body, and clothed nearly to the tarsal joint. *Feet* four-toed, three before and one behind; the former united by a membrane, the latter free. *Nails* falcate, their inner edges dilated.

SNOW GOOSE.

Anser hyperboreus, *Pall.*

L'Oie hyperborée, ou de neige.

THIS fine species of Goose inhabits all the regions of the Arctic circle, but more especially those portions appertaining to North America; it has also been said to inhabit the Antarctic circle, but this we find is not the case, its place being there supplied by another distinct species. From the northern portions of Russia and Lapland, where it is sparingly diffused, it regularly migrates to the eastern portions of Europe, and is occasionally found in Prussia and Austria, but never in Holland. The polar regions being its true and congenial habitat, it retires to those remote parts early in spring to perform the duties of incubating and rearing its young.

The eggs are of a yellowish white, of a regular ovate form, and somewhat larger than those of the Eider Duck.

Dr. Latham informs us that the Snow Goose is very numerous at Hudson's Bay; that it visits Severn River in May, and after having proceeded further north to breed, returns to "Severn Fort in the beginning of September, and remains to the middle of October, when they depart southward with their young, in flocks innumerable. At this time many thousands are killed by the inhabitants, who pluck them, and taking out the entrails, put their bodies into holes dug in the ground, covering them with earth, which, freezing above, keeps them perfectly sweet throughout the severe season; during which the inhabitants occasionally open one of these storehouses, when they find them sweet and good."

Its food consists of insects, rushes, and the roots of reeds and other vegetables, which, says Wilson, it tears " up from the marshes like hogs," and for which purpose its powerful serrated bill would seem to be expressly adapted: in autumn it feeds principally upon berries, especially those of the *Empetrum nigrum.* Like all the other vegetable feeders of the family, its flesh is very juicy, and forms an excellent article for the table.

The sexes are alike in colouring and may be thus described:

Forepart of the head as far as the eyes yellowish rust colour; the remainder of the plumage pure white, with the exception of the nine exterior quills, which have their bases and shafts white and the remainder black; bare space round the eye, bill, and feet rich reddish orange; nails of the mandibles blue; irides greyish brown.

The young exhibit a very striking difference from the adults, and have the head and upper part of the neck white; the remainder of the neck, breast, and upper part of the back purplish brown; all the feathers finely tipped with pale brown; wing-coverts, lower part of the back, and rump pale ash; primaries and secondaries black; tertials centred with black, and edged with light blue; tail-coverts white; tail blackish brown edged and tipped with white; belly and vent greyish; bill and feet light reddish purple.

We have figured an adult about one third less than the natural size.

GREY LAG WILD GOOSE.

Anser palustris. *(Flem.)*

E. Lear del.

GREY LAG WILD GOOSE.

Anser ferus, *Steph.*

L'Oie cendrée ou premiere.

NOTWITHSTANDING the variety of plumage which exists in our race of domesticated Geese, there is so striking a similitude in the form of the body, the shape and colouring of the bill, and other characters, as to leave no doubt in the minds of naturalists that they have descended from one common stock, of which the figure in our plate is a representative in its wild or natural state. The value of this bird as an article of food, and the various uses made of its feathers, are so well known to all our readers that it will be quite unnecessary for us to describe the management and rearing of the numerous domestic varieties, a subject so well understood by every one, and for a full account of which we refer our readers to the works of Pennant, &c. Although we learn from the testimony of older authors that this bird was once a permanent resident in the British Islands, it is now scarce, in consequence of its not being able to find a secure retreat where it may rear its young, the progress of cultivation and the drainage of the land compelling it to retire to more distant countries, where it may still breed unmolested.

The Grey Lag is known to inhabit all the extensive marshy districts throughout the temperate portions of Europe generally; its range northwards not extending further than the fifty-third degree of latitude, while southwards it extends to the northern portions of Africa, eastwardly to Persia, and, we believe, is generally dispersed over Asia Minor.

The Grey Lag assembles in flocks, and like the Bean Goose seeks the most open and wild districts, often descending upon fields of newly sprung wheat, which, with the blades of fine grasses, trefoil, and grain, constitute its food.

The nest is said to be placed among rushes, and is formed of a large quantity of various vegetable matters: the eggs, from six to twelve in number, of a sullied white.

The sexes are nearly alike in plumage.

Head and neck brown, tinged with grey; back, scapulars, and wing-coverts brown, tinged with ash grey, all the feathers being broadly margined with greyish white; lesser wing-coverts bluish grey; upper tail-coverts white; breast and belly greyish white, crossed with bars of a deeper tint; vent and under tail-coverts white; bill reddish orange, the nail greyish white; legs and feet dull red.

The Plate represents an adult male about two thirds of the natural size.

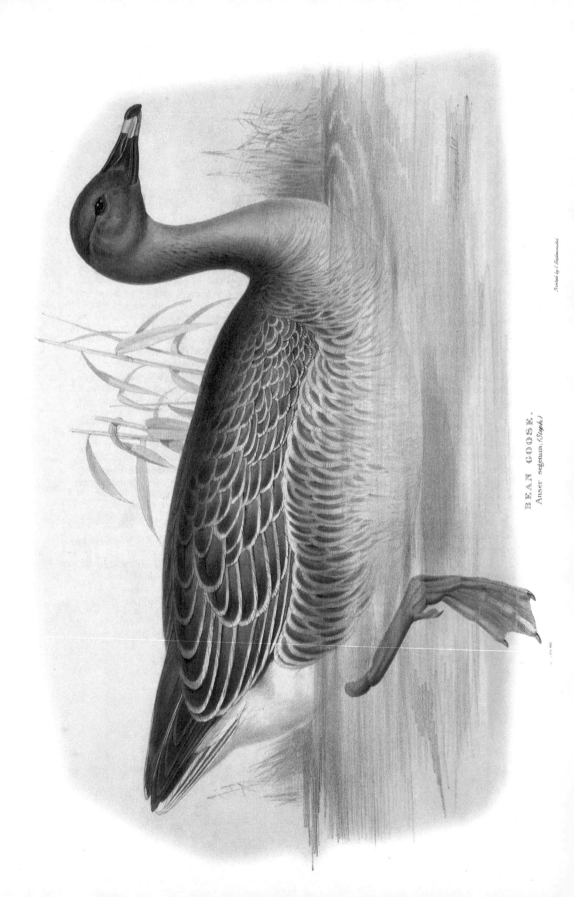

BEAN GOOSE.
Anser segetum. (Steph.)

Printed by C. Hullmandel

BEAN GOOSE.

Anser segetum, *Steph.*

L'Oie vulgaire ou sauvage.

In the temperate portions of Europe, and particularly on the British Islands, the Bean Goose is rather a winter visitor than a permanent resident: after passing the summer within the arctic circle, it migrates early in the autumn to more southern latitudes; hence in October and November considerable flocks arrive in the northern counties of England, and afterwards disperse themselves over the greater portion of our island. Like the other members of its family it is extremely shy and difficult of approach, cautiously avoiding danger by resorting to wild and open parts of the country. These birds feed principally by day, when they resort to stubble lands, and not unfrequently to grounds lately sown with peas, beans, and pulse, and they have been known to commit considerable havoc on young wheat. Extensive marshes and fenny districts also form a place of favourite resort, the large sheets of water which there abound affording them a safe retreat in cases of danger, and to which they frequently retire for the night. From the delicate quality of their food, the flesh of the Bean Goose is well flavoured, hence it is highly esteemed for the table, and numbers are annually sent to our markets, where they may be frequently observed exposed for sale, accompanied by their near ally the Grey Lag. On the approach of spring they commence their migrations northward; and notwithstanding the assertions made by some authors, that they remain and breed in our western and northern islands, we feel confident that the greater number make the high northern latitudes the place of general rendezvous.

The power of flight of the Bean Goose is very great, and when at a considerable elevation, with a favourable breeze, the rate at which it passes through the air has been estimated at from 60 to 80 miles an hour at least.

From the great similarity which exists between the Bean Goose and the Grey Lag, the two species have been often confounded; on a close examination, however, we cannot but observe the great difference which exists in the form and colour of their bills, as well as other peculiarities, which will be found in comparing the descriptions of the two birds. In size the Grey Lag has the advantage of the present species, although in this respect the male of the latter often exceeds in weight the female of the former. The black colour and diminutive bill of the Bean Goose, when opposed to the robust and flesh-coloured bill of the Grey Lag, forms perhaps the best and readiest mark of distinction between the two species.

The sexes are so nearly alike in the colour of their plumage that one description will serve for both; nor does there appear to be any perceptible change in their plumage at opposite seasons.

It is said to breed in low marshy situations, the female laying from eight to twelve white eggs.

The whole of the bill is black, with the exception of a band of pinky yellow (sometimes inclining to red) which surrounds both mandibles near the point; irides and orbits brown; top of the head and back of the neck brown, the latter having longitudinal furrows, giving this part the appearance of being marked with dark lines; the whole of the back, wings, flanks, and tail dark clove brown tinged with grey, each feather being tipped with white; breast and abdomen greyish brown; vent, under tail-coverts, and rump white; legs and webs orange.

The Plate represents an adult male about a third less than the natural size.

WHITE-FRONTED GOOSE.

Anser albifrons, (Bechst.)

WHITE-FRONTED GOOSE.

Anser albifrons, *Linn.*

L'Oie rieuse, ou à front blanc.

WE have not been able to ascertain with any degree of certainty whether the jet black markings which ornament the breast of this species are only assumed during summer, or whether they are confined to certain examples. While describing this bird in his "Manuel d'Ornithologie," we find M. Temminck equally at a loss with ourselves, for in a note appended to his description he states, that "after inspecting the plumage of this Goose, I suspect that the species moult twice in the year, and that in summer the belly and chest are of a deep black, while these parts in winter are pure white. I say only that case appears to me such, for it is by the naturalists in the North alone, who are able to observe this bird during the period of incubation, that this circumstance can be decided." The great numbers of this species brought to the London market during the first fortnight of January 1835, afforded us ample opportunities of examining a great variety of specimens; which examinations have led us to doubt if the irregular markings on the breast are ever assumed by many of these birds until they have attained a considerable age, there having been numerous mature individuals among those examined which had not the slightest trace of this peculiar feature, while in others it was slightly apparent; yet the majority of both sexes possessed the character in question, displayed in the most conspicuous manner. Some few birds of the year were observed, many of which had already the black feathers appearing in a slight degree; from which circumstance it is clear that a still further knowledge of the habits, changes, and economy of this fine Goose is required to render its history complete; and it is to be regretted that our island does not afford a retreat for so valuable a bird, in which it might breed and rear its young. It is just possible that the individuals examined by us at that early period of the year might not have commenced their spring moult; and if so, it is singular that others should have acquired so much of the black, which, according to M. Temminck's theory, is characteristic of the plumage of summer.

The summer residence of the White-fronted Goose is the high northern latitudes of both worlds: in these countries it finds a place of security wherein to incubate. It commences its migrations southwards early in the autumn, at which period great numbers pass into Holland, Germany, and France. In the British Islands it is tolerably abundant, particularly in the midland and southern counties, giving a decided preference to low marshes and fenny districts. Its food consists of aquatic vegetables of various kinds, small snails, &c. Its weight generally varies from five to seven pounds, and as an article for the table it is not surpassed by any of its tribe, its flesh being finely flavoured and tender.

Bill pale flesh-colour, nail pure white; forehead white; head, neck, and upper part of the chest greyish brown; breast, belly, and abdomen black marked with irregular bars and blotches of white; back brown, each feather being margined with greyish white; wing-coverts grey edged with white; quills dark greyish black; vent and upper tail-coverts white; middle tail-feathers grey, with white tips, the remainder becoming gradually lighter, to the outer ones, which are wholly white; legs and toes orange; claws whitish.

The Plate represents an adult and a young bird of the year about three fourths of the natural size.

BERNICLE GOOSE.
Anser leucopsis; (Bechst.)

BERNICLE GOOSE.

Anser leucopsis, *Bechst.*

L'Oie Bernache.

Iт is only during the months of autumn and winter that the British Islands are visited by the Bernicle Goose, the extreme cold of the northern latitudes, where it sojourned during the summer season, having driven it southward into climates where its food is still accessible. The portion of our island in which it is most abundant is along the whole of the western coast from north to south. In Lancashire it appears especially abundant: the North and West of Ireland is also visited by it in large flocks. On the Continent Holland, Germany, and France offer an extent of coast and inland meres and marshes highly acceptable to the Bernicle, to which localities it resorts in great numbers. It is decidedly one of the handsomest and most elegant of the Geese that sojourn in the British dominions, and when domesticated forms a graceful ornament to our aviaries. When wild it is extremely shy and wary, so much so that it cannot be approached without the utmost circumspection. Its food consists of various aquatic and terrestrial vegetables, seeds, and grain. It breeds in the regions of the arctic circle, but we have no correct information as to the description of its eggs, or its peculiar habits of nidification, in which, however, we conceive it agrees with the rest of its congeners.

The sexes offer so little difference in the colouring of their plumage that one description will serve for both.

The adult has the forehead, cheeks, and throat yellowish white; a narrow black mark passes from the bill to the eye; the top of the head, neck, and chest black; the upper surface fine blueish grey, the tip of each feather edged with brownish black and a margin of greyish white beyond; primaries greyish black; upper tail-coverts white; tail black; the whole of the under surface silvery white; flanks strongly marked with grey in waved bars; feet and bill black; irides dark brown.

The young are easily distinguished from the adults, by the light colouring of the face being more clouded with black, and by the general plumage being less pure and decided.

The Plate represents an adult about three fourths of the natural size.

RED-BREASTED GOOSE.
Anser ruficollis. (Pall.)

RED-BREASTED GOOSE.

Anser ruficollis, *Pall.*

L'Oie à Cou roux.

WE regret that we are unable to give any detailed account of this beautiful Goose. Only four or five instances are on record of its having been captured in the British Islands; and its occurrence on the European continent appears to be equally rare, except in the most north-eastern portions, where it is rather more plentiful. The countries to which it habitually resorts are doubtless the extreme northern parts of Asia and Siberia, its migrations in summer extending to the shores of the Frozen Ocean, where it breeds and rears its young. Extraordinarily severe seasons or other unusual circumstances, driving it out of its usual course, are in all probability the causes of its appearing occasionally in this country, and in other temperate portions of the globe. The first example captured in England was taken near London in 1776, passed into the hands of Mr. Tunstall, and is now in the Museum at Newcastle-upon-Tyne; another was captured alive near Wycliffe, and was kept in confinement for some years by the gentleman above mentioned; a third was killed near Berwick-upon Tweed, and formed a part of Mr. Bullock's celebrated collection; and others, Mr. Stephens informs us, were killed in the severe winter of 1813 in Cambridgeshire.

In its habits, disposition, and food it doubtless offers a strict resemblance to the other members of its genus: that it feeds on vegetables is pretty certain, from the circumstance of its flesh being free from any fishy taste and in great esteem for the table.

We are not aware whether the sexes exhibit any difference in their colouring, but judging from analogy we should conceive that they do not vary much, if any.

Forehead, top of the head, stripe down the back of the neck, chin, throat, and a band extending upwards to the eye, black; on each side of the head a patch of reddish brown surrounded by a stripe of white, which is extended down the sides of the neck, and separates the black stripe down the back of the neck from the reddish brown of the lower part of the neck and breast, which latter colour is margined with black, to which succeeds a stripe of white; upper surface, fore part of the belly, wings, and tail black; hinder part of the belly, vent, thighs, upper and under tail-coverts white; greater wing-coverts margined with white; bill and legs blackish brown.

We have figured a male somewhat less than the natural size.

BRENT GOOSE.

Anser Brenta. /Flem./

E. Lear del.

BRENT GOOSE.

Anser Brenta, *Flem.*

L'Oie cravant.

Tʜɪs well-known species is the least of the European Wild Geese, and is one of our winter visitants, at which period it resorts in great numbers to the inlets of the sea and the bays around our coast. Being driven from the icy regions of the polar circle by the approach of inclement weather, it not only visits the shores of the British Islands, but appears to radiate in every direction, spreading itself over almost all the maritime portions of Europe, Asia, and America. "Upon the Northumbriam coast," says Mr. Selby, "a very large body of these birds annually resorts to the extensive muddy and sandy flats that lie between the mainland and Holy Island, and which are covered by every flow of the tide. In this locality tolerably sized flocks usually make their appearance in the early part of October, which are increased by the repeated arrival of others till the beginning of November, at which time the equatorial movement of the species in this latitude seems to be completed. This part of the coast appears to have been a favourite resort of these birds from time immemorial, where they have always received the name of *Ware Geese*, given to them, without doubt, in consequence of their food consisting entirely of marine vegetables. This I have frequently verified by dissection ; finding the gizzard filled with the leaves and stems of a species of grass that grows abundantly in the shallow pools left by the tide, and with the remains of the fronds of different *Algæ*, particularly of one which seems to be the *Laver* (*Ulva latissima*). These were mixed with a considerable quantity of sharp sand, but without any portion of animal or shelly matter, although Wilson states that they feed occasionally upon small univalve and bivalve mollusca. In this haunt they remain till the end of February, when they migrate in successive flocks, as the individuals happen to be influenced by the season, and before April the whole have disappeared. When feeding (which they do at ebb tide) or moving from one place to another, they keep up a continual hoarse cackling, or, as it is termed, *honking* noise, which can be heard at a great distance."

The Brent Goose is always extremely shy and watchful, and can only be approached by the sportsman concealing his person. This wariness has induced those who procure these birds for the market to resort to various contrivances to effect their object, for an account of which we beg to refer our readers to Colonel Hawker's 'Instructions to young Sportsmen,' an amusing treatise, where a full description is given of this kind of sporting.

They breed and rear their young in the security of high northern latitudes ; the nest being formed of various vegetable materials, and the eggs, which are white, being ten or twelve in number.

The male has the head, neck, and upper part of the breast black ; on each side of the neck a patch of white ; back, scapulars, and wing-coverts brown, each feather being margined with paler brown ; under surface dark grey, each feather margined with paler grey ; vent and upper and under tail-coverts white ; lower part of the back, the rump, quills, and tail black ; bill black ; irides brown ; legs and feet brownish black.

The female resembles the male in colour, but is not quite so large.

Our Plate represents an adult male rather less than the natural size.

EGYPTIAN GOOSE.
Chenalopex Ægyptiaca. *(Steph.)*

Genus CHENALOPEX, *Antiq.*

GEN. CHAR. *Beak* as long as the head, slender, straight, its tip rounded, the margin laminated; the upper mandible curved, its tip hooked; the lower mandible flat. *Nostrils*
placed at the basal portion of the beak. *Wings* armed with spurs. *Legs* placed in the
equilibrium of the body, four-toed; the anterior toes entirely webbed, the hinder one
simple; tarsi somewhat elongated.

EGYPTIAN GOOSE.

Chenalopex Egyptiaca, *Steph.*

L'Oie d'Egypte.

ON comparing the present species with the other members of its family, it will be found to differ in form from
every one of them, on which account it has been formed by Mr. Stephens into a distinct genus, to which he
has been induced to give the above generic title, in consequence of this bird being in the opinion of M. Geof-
froy St. Hilaire the *Chenalopex* or Vulpanser of the ancients.

In figuring this fine species of Goose as a member of the European Fauna, we are not instigated by the
occurrence of numerous half-reclaimed individuals which are yearly shot in our island, but from the circumstance of its occasionally visiting the southern parts of the Continent from its native country Africa. M. Temminck particularly mentions the island of Sicily as one of the places frequented by it. This is the species
which would appear to have been held in great veneration by the ancient Egyptians, as we frequently find a
figure of it among the stupendous works of that celebrated people. It is abundant on the banks of the Nile,
and is distributed over the whole of the vast continent of Africa.

It readily breeds in confinement, and forms a beautiful and interesting addition to the menagerie.

The sexes are alike in plumage, but the female is somewhat smaller in size, and has the whole of the
markings less decided than in the male.

Feathers immediately behind the base of the bill, a narrow line running from the upper angle of the gape
to the eye, and a large patch surrounding the eye, rich chestnut; sides of the face, crown of the head, and
the fore part of the neck buffy white, gradually passing on the back of the neck into rufous brown; this
reddish tinge also predominates on the lower part of the neck, and forms a faint collar; upper part of the
back light chestnut brown, transversely rayed with very minute and irregular lines of blackish brown; centre
of the back and upper part of the scapularies dark reddish brown, minutely rayed with irregular transverse
lines of blackish brown and grey; lower part of the scapularies and tertiaries rich reddish chestnut; lesser
wing-coverts pure white with the exception of the posterior row of feathers, which are crossed with a strongly
defined mark of black near their extremities, forming a narrow band across the wing; primaries, lower part
of the back, rump, and tail black; secondaries rich glossy green, with purple reflections; on the centre of
the breast a large irregular patch of deep rich chestnut; all the remainder of the under surface from the
collar to the thighs pale buff, transversely rayed with very minute and irregular lines of blackish brown; vent
and under tail-coverts rich buff; upper mandible margined all round with brown, the centre being reddish
flesh colour; legs and feet reddish flesh colour; irides orange.

We have figured an adult male about one third less than the natural size.

DOMESTIC SWAN.
Cygnus mansuetus. (Gmel.)

Genus CYGNUS.

Gen. Char. *Beak* equally wide throughout its length, much higher than broad at the base, where it is swollen or tuberculated ; depressed towards the tip ; nail of the upper mandible deflected, and covering that of the lower, which is flat. Both mandibles laminato-dentate, with the lamellæ placed transversely, and nearly hidden from view when the beak is closed. *Nostrils* oblong, lateral, placed near the middle of the beak. *Wings* large and long. *Legs* short; feet four-toed, three before, one behind; the front toes entirely webbed, the hind toe small and free.

DOMESTIC SWAN.

Cygnus mansuetus, *Gmel.*

Le Cygne.

The Domestic Swan, the stately ornament of our lakes and rivers, is too well known to render much description necessary. The ease and grace with which this bird ploughs its course along the rippled surface of the water, has raised it to that high rank in general estimation to which its extreme beauty and peaceful habits so fully entitle it.

The Swan is one of the largest of our indigenous birds, frequently weighing from twenty to twenty-five pounds. The bill is orange colour, the base and cere reaching to the eye, black, and surmounted with a fleshy knob of the same black colour ; the legs and feet are also black ; all the other parts in the adult bird are of a pure and spotless white. The first plumage of the Cygnet, or young Swan, is of a dull brownish ash colour, afterwards varied with white ; but the young birds do not attain their pure and perfect white appearance till their second year, and are incapable of breeding before the third year. The parent birds drive away from them the brood of the previous year as soon as the revolving seasons again produce the period of incubation. At this time the male assumes an appearance of boldness and contempt of danger which plainly indicate the change in his habits which the season has produced. The male may be distinguished from the female by his thicker neck and his wider and shorter body ; and the female appears to swim deeper in the water.

The female lays six or seven long oval-shaped eggs, of a greenish grey colour, and sits about forty-five days. During this extended period, the male keeps watch at a short distance from her nest ; and when the young brood are produced, and take to the water, he is incessant in his care and guardianship, and boldly advances to repel the intruder upon every appearance of danger.

Formerly young birds of the year were in great request as an article of food, and were frequently served up as a choice dish on great occasions ; even now young Swans, intended for the table, are occasionally to be seen, in their grey plumage, at the shops of our London poulterers.

Although a few Swans may be observed on most of the lakes which ornament the parks and grounds of the nobility and others, they are nowhere very numerous, if we except the swannery of the Earl of Ilchester, at Abbotsbury in Dorsetshire, where a large stock has been maintained for many years. The various parts of aquatic plants are the natural food of these birds, in search of which they examine all the shallow parts of the water they inhabit, and are able to keep their head below the surface for a considerable length of time, but are never seen to dive. In confinement they feed readily on grain, for the comminution of which their large and powerful gizzard seems well adapted.

The voice of the tame Swan is feeble, plaintive, and not unmusical ; but this bird does not possess internally that convoluted structure of trachea which has made the examination of the various wild species an object of so much interest, and which we shall have occasion to notice more particularly when describing the Hooper and Bewick's Swan. Our Domestic Swan is said to exist in a wild state in Russia and Siberia ; but we must not omit to mention, that a species called the Polish Swan has lately been introduced to this country, which, compared with the subject of our present Plate, exhibits a slight difference in the distribution of the colours on the beak, and in the situation of the nostrils : the legs and feet are of a greyish ash colour, and the young birds are said to be white from the egg, never afterwards assuming any of that ash colour which distinguishes till their second year the Cygnets of other white Swans.

WHISTLING SWAN.

Cygnus ferus, (Rey).

C. Gould del

WHISTLING SWAN, OR HOOPER.

Cygnus ferus, *Ray.*

La Cygne à bec jaune ou sauvage.

WE refer to the present species of wild swan by the name of Whistling Swan, or Hooper, in order to distinguish it from two other species of wild swans which have recently been added to this genus, one of which, an occasional visitor to England, Ireland, and other parts of Europe, we have figured in this work. The term Hooper has the advantage of referring to a peculiar character of the voice in the present bird, which is as yet considered to be specific: its usual call-note resembles the sound of the word *hoop*, loudly and harshly uttered several times in succession.

The Hooper has usually been considered an inhabitant of North America, but anatomical examination of the two species of wild swans most numerous there proves that they are both distinct from the Hooper; and it will probably be found that this last-named species is exclusively confined to the northern parts of Europe and Asia.

The Hooper is only a winter visitor in England or in the southern countries of the European continent, and the number seen there during that season of the year generally bears some proportion to the degree of severity in the weather. During long-continued frosts large flocks are not uncommon, and our markets afford numerous examples; but in mild winters few are obtained or even seen. The summer residence of the Hooper is within the Arctic circle, in Iceland, Scandinavia, and the most northern countries of Europe. Formerly a few pairs were known to rear their young among the islands of Shetland and Orkney, and even in Sutherlandshire. In a half-domesticated state, with pinioned wings, the Hooper breeds about the lakes and islands in the parks of some English noblemen, but it does not, in such situations, associate much with the Domestic Swan, which is the more usual monarch of ornamental waters.

The food of the Hooper are aquatic plants and insects, feeding in shallow water: it makes a large nest on the ground, collecting leaves, rushes, or flags, and lays six or seven whitish eggs, which are tinged with a yellowish green; the length of the egg four inches, the breadth two inches and three quarters. The parent bird sits six weeks: the young are at first of a uniform dark grey, acquiring a white plumage by slow degrees about the time of completing their second autumn moult, previous to which the dark anterior part of the beak is not decidedly black; the base of the beak and the cere are more of a fleshy tint than yellow, and the legs are also lighter in colour than those of the old birds.

The adult female only differs from the male in being smaller, and the neck is more slender.

In the adult male the plumage is perfectly white, if we except an occasional tinge of buff-colour on the top of the head; the beak black, the base and cere yellowish orange, this colour extending forwards along the edges of the upper mandible as far as the line of the most anterior part of the nostrils, and posteriorly surrounding the eyes; irides brown; the legs and feet black; the whole length of the bird about five feet; the breadth with extended wings nearly eight feet.

The papers of Dr. Latham and Mr. Yarrell in the Transactions of the Linnean Society of London, on the organs of voice in birds, contain descriptions and figures of internal peculiarities by which the species of Swans most likely to be confounded may be readily distinguished.

The Plate represents an adult about one third of the natural size.

BEWICK'S SWAN.
Cygnus Bewickii. (Yarr.)

BEWICK'S SWAN.

Cygnus Bewickii, *Yarr.*

Le Cygne de Bewick.

THIS fine addition to the ornithology of Europe and Great Britain was made known at the commencement of the winter of 1829, although, as has since appeared, various specimens were before that time preserved in different collections; but the characters, principally internal, by which this new species is distinguished from the Hooper had not been ascertained. In a paper by Mr. Yarrell, read at the Linnean Society, the specific peculiarities belonging to the bony structure of this Swan proved satisfactorily that it was distinct from the Hooper, and the name of *Cygnus Bewickii* was proposed for it.

The appearance of this species in England seems to depend on the degree of severity of the winter, and, comparatively, but few have been seen here since the season of 1829–30. It is probably an inhabitant of the northern portions of the continents of Europe, Asia, and America.

Dr. Richardson in his 'Fauna Boreali-Americana' says, "This swan breeds on the sea-coast within the arctic circle, and is seen in the interior of the fur-countries on its passage only. It makes its appearance amongst the latest of the migratory birds in the spring, while the Trumpeter Swans are, with the exception of the Eagles, the earliest. It winters, according to Lewis and Clarke, near the mouth of the Columbia." Captain Franklin, in the Journal of his second expedition to the Arctic regions, when residing at the station on the Great Bear Lake during the winter of 1827, remarks: "We welcomed the appearance of two large-sized swans (Trumpeters) on the 15th of April as the harbingers of spring; and on the 20th of May, the small-sized swans (*C. Bewickii*) were seen, which the traders considered the last of the migratory birds." Captain Lyon describes the nest of Bewick's Swan as built of moss-peat, nearly six feet long, by four and three-quarters wide, and two feet high exteriorly; the cavity a foot and a half in diameter. The eggs were brownish white, slightly clouded with a darker tint.

In size Bewick's Swan is one third smaller than the Hooper at the same age. The plumage is first grey, afterwards white tinged with rust colour on the top of the head and on the under surface of the belly, and ultimately pure white. The beak is black at the point, and orange yellow at the base in the males; this last colour appears first on the sides of the upper mandible, and afterwards covers the upper surface in front of the forehead to the extent of three quarters of an inch, receding from thence by a convex line to the lower edge of the mandible at the gape; the nostrils are oblong; the irides orange yellow; the wings have the second and third primaries the longest and equal, the first and fourth half an inch shorter than the second and third, and also equal; the tail consists of twenty feathers, graduated, cuneiform; the legs, toes, and claws black. The base of the beak in females is lemon yellow. The food of this species is similar to that of the Hooper.

The internal characters which distinguish the two Wild Swans found occasionally in England are as follows:

In the Hooper the tube of the trachea, or windpipe, is not uniform in size throughout its length, and that portion of it which is confined within the keel of the breastbone never departs from a vertical position at any age, nor is there any excavation in the sternum itself. The bronchial tubes are invariably long.

In Bewick's Swan the tube of the windpipe is of equal diameter throughout its length, and when arrived at the end of the keel of the sternum it inclines upward, and passes into a horizontal cavity destined to receive it, caused by the separation of the two horizontal plates of bone forming the posterior flattened portion of the breastbone. The bronchial tubes are short. Descriptions and figures of the organs of voice of the Wild Swans will be found in the Linnean Transactions already referred to.

The whole length of Bewick's Swan is three feet ten inches.

We have figured a male about one third of the natural size.

COMMON SHIELDRAKE.
Anas tadorna, (Linn.)
Tadorna vulpanser, (Flem.)

Genus TADORNA.

Gen. Char. *Bill* shorter than the head, higher than broad at the base, depressed or concave in the middle, the tip flattened and turning upwards, nearly of the same breadth throughout; *dertrum,* or *nail,* abruptly hooked; upper mandible laterally grooved near the tip; under mandible much narrower than the upper one, and, when closed, hidden by the deflected *tomia* of the upper; both mandibles having prominent transverse lamellæ. *Nasal fosse* near the base of the bill; *nostrils* oval, lateral, pervious. *Wings* of mean length, acute, tuberculated, with the second quill-feather the longest. *Legs* of mean length, with the tibiæ naked for a short space above the tarsal joint. *Tarsus* rather longer than the middle toe. *Toes* four, three before and one behind; the front ones rather short and entirely webbed; hind toe barely touching the ground with the tip of the nail. *Claws* slightly hooked, the inner edge of the middle one being dilated.

COMMON SHIELDRAKE.

Anas tadorna, *Linn.*

Tadorna vulpanser, *Flem.*

Le Canard tadorne.

With few exceptions, the Common Shieldrake may be considered one of the most beautiful of its race; there is certainly no European species which exceeds it in graceful motion or simplicity of colouring; and when domesticated, it adds great beauty and ornament to our lakes and sheets of water, where, notwithstanding it is a native of the sea shore, it lives and thrives without any difficulty, sailing about with its mate, which closely resembles it in colouring, as if to display its symmetry and the fine contrast of its tints to the best advantage.

It is distributed throughout the whole of Europe, and is moreover indigenous to the British Isles, breeding upon some parts of our coast in considerable abundance. The situations it chooses for the purpose of nidification are both singular and novel: these are no other than the deserted burrows of the rabbit, which are abundantly scattered over the sand-hills adjacent to the shore on several parts of the coast; and here the female constructs a nest, at the distance of many feet from the entrance, consisting of dried grasses and other vegetable materials, and lined with down from its own breast: the eggs are pure white, and from twelve to sixteen in number. Like many other birds, the male and female sit alternately; and the young so soon as they are hatched are conducted, or, as it is said, frequently carried in the bills of the parents, to the sea, which is the congenial element of this species, as they merely retire inland to the salt marshes and saline lakes for the purpose of feeding.

If we attend to the form of this bird, we cannot fail to observe the situation which it fills in the family to which it belongs: its general characters indicate it as belonging to the true Ducks, while its lengthened tarsus and elevated hind toe, together with its mode of progression on the ground, denote an affinity to the Geese; points which have led to the formation of the genus *Tadorna*, of which this and the *Anas rutila* form the only European examples.

The note of the Common Shieldrake is shrill and whistling. The fleshy tubercle on the top of the upper mandible acquires in the spring a more heightened and brilliant tinge of crimson than it possesses at the other seasons of the year. Its food consists of insects, shelled mollusca, crustacea, and marine plants.

The male and female, as in the true Geese, offer but little difference of plumage; the latter is, however, somewhat smaller in size, and her colours are more obscure. M. Temminck states that it is found in all the northern and western countries of Europe, along the borders of the sea, being abundant in Holland and France, and accidentally appearing in the rivers of Germany and other parts of the Continent.

The whole of the head and upper parts of the neck glossy black; the lower part of the neck, the shoulders, sides of the abdomen, back, tail, upper and under tail-coverts, white; the tail being tipped with black, which colour runs down the middle of the belly, and covers the greater portion of the scapulars and greater quill-feathers; a broad band of chestnut encircles the breast and upper part of the back; speculum of the wing brilliant green; beak bright orange-red; tarsi and feet flesh-colour.

The young have the forehead, fore part of the neck, and under parts, inclining to white.

The Plate represents the male three fourths of the natural size.

RUDDY SHIELDRAKE.
Tadorna rutila. *(Steph.)*
Anas Casarka. *(Linn.)*

Drawn from Nature & on stone by J & E Gould.

Printed by C Hullmandel.

RUDDY SHIELDRAKE.

Tadorna rutila, *Steph.*

Anas Casarka, *Linn.*

Le Canard Kasarka.

WHILE we follow Messrs. Stephens and Selby in placing this magnificent Duck in the genus *Tadorna*, we are not satisfied that the situation assigned to it is perfectly correct; it appears to us to constitute the type of a distinct form, of which the *Anas Tadornoïdes*, Jard., of New South Wales will form a second example. We think it approaches nearer to the true Geese than to the Shieldrakes, inasmuch as it possesses a rounder form of body, stands higher upon the legs, and has a shorter bill, better adapted for grazing or nibbling grasses and aquatic vegetables, which constitute its principal food.

As a European bird the *Tadorna rutila* may be considered as one of the rarest, and more particularly so as a British species, not more than two or three instances of its occurrence in our island being on record; one of which, as stated by Mr. Fox in his Synopsis of the Newcastle Museum, was killed at Bryanstone, near Blandford, in Dorsetshire, the seat of Mr. Portman, in the severe winter of 1776. On the Continent it inhabits Russia and other eastern districts, and is occasionally met with in Austria and Hungary. It is dispersed over a great part of Asia, and it would appear to be also an inhabitant of Africa, specimens brought from thence offering no differences from those individuals killed in Europe.

The Ruddy Shieldrake is rarely found on the sea-coast, but dwells and breeds upon the borders of large rivers, in situations similar to those selected by the common species, and lays from eight to ten white eggs.

Its food consists of grasses, aquatic plants, and insects.

The whole of the head and neck pale ochreous yellow, becoming gradually darker until it meets a collar of deep black glossed with green, which surrounds the neck; breast, back, scapulars, and the whole of the under-surface rich chestnut red; lesser and middle wing-coverts yellowish white; secondaries purple glossed with green; quills black; lower part of the back, upper tail-coverts, and tail dull black; bill, legs, and feet black.

The female is destitute of the black collar, is less brilliant in colour, and has the feathers of the back finely speckled with grey.

We have figured an adult male about two thirds of the natural size.

WIGEON.

Anas Penelope; (Linn.)

Mareca Penelope; (Selby.)

Genus MARECA, *Steph*.

GEN. CHAR. *Bill* shorter than the head, higher than broad at the base, straight from before the
nostrils, flattened and narrowing towards the tip, which is armed with a hooked nail ;
mandibles laminate-dentate, with the points of the laminæ of the upper mandible slightly
projecting in the centre of the bill beyond the margins. *Nostrils* lateral, placed near the
base of the bill, small, oval, pervious. *Wings* acuminate. *Tail* wedge-shaped, consisting
of fourteen feathers, acute. *Hind toe* small, having a narrow web.

WIDGEON.

Mareca Penelope, *Selby*.

Anas Penelope, *Linn*.

Le Canard siffleur.

OF the many species of the Duck tribe which visit this country annually, though not indigenous to our
islands, the Widgeon is one which is especially abundant during the autumn and winter months of the year,
associating in flocks upon our meres and inland lakes, as well as the larger streams and rivers, whence, if the
weather is unusually severe, so as to prevent its obtaining its favourite food, it passes to the open coasts,
particularly such as are bordered by long swampy tracts of land. During the time they remain with us,
multitudes are annually taken in decoys, while not a few fall a sacrifice to the gun, their flesh, which is both
delicate and savoury, being highly esteemed for the table. They are also found in great abundance in the
lowlands of France, Germany, and Holland, as well as in all other similar portions of the Continent.

The Widgeon may be considered as strictly a vegetable feeder ; and in the manner of taking its food it
differs much from the generality of ducks, in as much as it usually feeds near the edge of the water, nibbling
or biting off the tender blades of grass and other herbage.

In the month of March the multitudes which have been sojourning in our southern latitudes wing their
way northwards, where they pass their summer, incubate, and bring up their young. Though it is not
improbable that stragglers may remain and breed in our latitudes, still it must be confessed that those retained
as prisoners, under the most favourable circumstances, have scarcely, if ever, been known to breed ; the usual
changes of plumage which are so conspicuous in this species of duck are, nevertheless, regularly exhibited,
the male losing his variegated tints towards the end of summer, and becoming very similar to the female.

The eggs are said to be eight or ten in number, and of a dull greyish green.

The figures in the Plate represent the male and female in the plumage of winter and spring, which may
be thus described.

Male : Top of the head pale buff ; cheeks and neck deep chestnut ; the ear-coverts spotted with black ;
chest delicate vinous grey ; the upper surface generally, and the flanks, beautiful grey minutely barred
with fine zigzag lines of black ; under surface and centre of the wing white ; speculum green ; bill and
legs blueish lead colour, the former tipped with black.

The female is of a dusky reddish brown ; the head and neck thickly spotted with dark brown, each feather
having a lighter margin, which produces a scaly appearance ; under surface white ; bill blackish brown.

The figures are rather less than the natural size.

SHOVELLER DUCK.
Anas clypeata. *(Linn.)*
Rynchapsis clypeata. *(Leach.)*

Drawn from Nature & on Stone by J. & E. Gould.

Printed by C. Hullmandel.

Genus RHYNCHASPIS, *Leach.*

GEN. CHAR. *Beak* long, its base unarmed, semi-cylindric, the tip dilated, somewhat spoon-shaped, with a small incurved nail; the sides of the mandibles with pectinated lamellæ. *Nostrils* medial, oval, basal. *Tail* short, simple, furnished mostly with fourteen feathers.

SHOVELLER DUCK.

Rhynchaspis clypeata, *Steph.*

Anas clypeata, *Linn.*

Le Canard Souchet, ou le Rouge.

THE singular spoon-shaped bill bordered with numerous delicate laminæ, which characterizes several species of this interesting family, has been considered of sufficient importance to constitute the distinctive characters of a group, to which the name of *Rhynchaspis* has been applied by Dr. Leach, and that of *Spathulea* by Dr. Fleming, while Mr. Swainson retains the title of *Anas* to this group, as from the peculiar structure of its beak, he considers the Shoveller to be the type of the true grass-feeding ducks: we have restricted the term *Anas*, however, to the group comprehending the Common Wild Duck; and having been so employed by ourselves, and the term *Spathulea* having scarcely been adopted, we prefer the generic title of *Rhynchaspis*.

This group consists of several species, which are almost universally though sparingly distributed. The range of the present species extends throughout the temperate portions of Europe, the northern regions of Africa, and nearly the whole of India: in our island it appears to be somewhat limited; doubtless a few breed annually in our marshes, and Mr. Selby informs us that he has a male in his collection killed in the month of July, at which period it undergoes that transition of plumage which assimilates it to the female, a change the utility of which has not as yet been philosophically explained. As we have observed that this change is common to the males of those species that more especially breed in marshes, among reeds, &c., and as it generally takes place at the period of incubation, may it not serve as a protection to the species by rendering the fostering parent less conspicuous at this critical period than he would be were he to retain the gay nuptial dress, which would present so strong a contrast to the sombre-tinted vegetation among which it is necessary for him to remain, until the young are able to provide for themselves?

It is said to prefer lakes and inland waters to the sea and saline marshes, a circumstance to be accounted for by the peculiar nature of its food, which consists of the larvæ of insects, and freshwater vegetables, such as grasses and chickweed: from this kind of food its flesh, as might be expected, is both delicate and tender, and in high esteem for the table.

Its mode of nidification is very like that of the Common Wild Duck, the nest being constructed among coarse herbage in the central parts of marshes, and the eggs, being from ten to twelve in number, of a pale green colour.

The sexes differ considerably in their colouring, the male being adorned in spring and summer with a rich and delicate plumage; while the female is of a more uniform and sombre tint.

The male has the head and upper part of the neck deep brown glossed with green; lower part of the neck, breast, scapulars, and sides of the rump white; back blackish brown, each feather margined with grey and tinged with green; lesser wing-coverts and outer webs of some of the scapularies greyish blue; tips of the larger coverts white forming a bar across the wing; speculum rich green; tertials rich purplish black with a streak of white down the centre; middle tail-feathers brown edged with white, outer ones entirely white; upper and under tail-coverts black tinged with green; under surface yellowish brown with zigzag lines of black upon the flanks and vent; bill blackish brown.

The female has the whole of the upper surface deep brown, each feather barred and margined with reddish white.

We have figured a male and female, rather less than the natural size.

COMMON WILD DUCK.
Anas Boschas. (*Linn.*)

Drawn from Nature & on Stone by J & E Gould.

Printed by C Hullmandel.

Genus ANAS, *Linn.*

GEN. CHAR. *Bill* longer than the head, depressed through its whole length, broad, straight from before the nostrils to the tip, nearly equal in breadth throughout; mandibles dentato-laminate, with the laminæ of the upper mandible scarcely projecting beyond the margin. *Nostrils* lateral, oval, situated near the base of the bill. *Wings* of mean length, acuminate. *Tail* short, slightly wedge-shaped; the middle feathers curling upwards in some species. *Feet* with four toes, three before and one behind; the front ones webbed, the hind toe small and free.

COMMON WILD DUCK.

Anas Boschas, *Linn.*

Le Canard ordinaire.

THE circumstances attending the domestication of the Duck, like that of many other reclaimed animals, are buried in obscurity; and it is impossible to decide whether the attention of man was directed to it in conse-quence of the superiority of its flesh as an article of food, or whether of all the Duck tribe he found it most naturally inclined to submit to the arts of domestication. It is almost unnecessary for us to state that the present well-known species is the origin of our many domestic varieties.

The range of the Common Wild Duck extends over the whole of the temperate portion of the globe; and although we believe it is scarcely ever found in a wild state south of the equator, its extreme limits approach within a few degrees of the meridian. It is dispersed throughout this vast extent of country, and everywhere shows the same instinct, and the same disposition to become domestic and familiar. In our own island and the adjacent parts of the Continent, numbers remain to breed wherever they can find congenial situations; these numbers are greatly augmented in spring and autumn by an influx of visitors on their journey from north to south and back again. Great quantities proceed to the northern regions, where they continue in greater safety among the vast morasses of those countries. From the nature of its food, which consists almost exclusively of vegetables, its flesh furnishes a wholesome and nutritious diet, and is peculiarly tender and well flavoured. In the districts around its breeding-haunts, the young, before their primaries are fully grown, are known by the name of flappers, and from the richness of their flesh are in great requisition. So much has already been written respecting the wholesale mode of capturing the Wild Duck in decoys in the counties of Lincolnshire and Cambridgeshire, that it would be superfluous to say anything more on the subject.

In this country the Common Wild Duck commences breeding early in spring, pairing in the months of February and March, and selecting a secluded spot near the water's edge, where the female deposits her bluish white eggs and rears her progeny. After the young are able to shift for themselves, the parents separate from them and congregate in distinct flocks, and it is asserted that the sexes form separate bands. The young males do not attain their full plumage until the following spring.

The colouring of the adult male, or mallard, is peculiarly elegant.

The whole of the head and half the neck are of a deep metallic green; the middle of the neck is encircled by a ring of white; the chest is very deep chestnut; the centre of the back is brown, each feather having a lighter margin; the scapularies and flanks are greyish white, beautifully barred with fine zigzag pencillings of black; shoulders greyish brown; speculum rich changeable purplish green passing into velvety black, bounded both before and behind by bands of white; quills dark brown; rump and upper tail-coverts greenish black, the two longest or middle tail-feathers curling upwards, tail-feathers greyish white; under tail-coverts greyish black; bill olive yellow; legs orange.

The general plumage of the female is tawny brown, numerously marked about the head and neck with dusky spots; the feathers of the back, sides, and under surface having their centres of a deeper tint; the speculum of the wing resembles that of the male but occupies a smaller space.

The Plate represents a male and female rather less than the natural size.

COMMON TEAL.

Anas crecca, (Linn.)

Querquedula crecca. (Steph.)

Printed by C. Hullmandel.

Genus QUERQUEDULA, *Ray.*

GEN. CHAR. *Bill* as long as the head, elevated at the base, straight, semicylindrical, nearly of equal breadth throughout; tip obtuse, with the dertrum, or nail, small and hooked; mandibles laminated, and having the laminæ almost entirely concealed by the deflected margins of the upper mandible; *nasal fosse* small, lateral, near to the culmen of the bill. *Nostrils* oval, pervious. *Wings* acute, with the first and second quills of nearly equal length. *Tail* wedge-shaped, with the two middle feathers more or less elongated, and acute. *Legs* having the tarsus rather shorter than the middle toe. *Feet* with four toes, three before and one behind; the front ones webbed; the hind toe small and free.

COMMON TEAL.

Anas Crecca, *Linn.*

Querquedula Crecca, *Steph.*

La Petite Sarcelle.

THIS elegant little Duck, one of the smallest of the *Anatidæ*, is widely distributed over the Old World. It is abundant on the range of the Himalaya, whence we have received many examples, the collections brought home by Colonel Sykes and Major Franklin, the former from the western ghauts of India, and the latter from the plains intermediate between Calcutta and the Nepaul hills, affording us examples which, on comparison, are found to be strictly identical with our European birds, as are also specimens from Africa. M. Temminck names Northern America as among its native localities; but from this opinion we are inclined to dissent, for the American examples may always be distinguished by a white crescent-shaped band on each side of the chest near the shoulders. This, together with the absence of the white tertial feather, will, we think, constitute fair grounds for a genuine specific distinction.

In the British Islands, though it breeds in the northern districts, its numbers are greatly augmented in winter by visiters from the high latitudes of the Continent, which spread themselves over the marshy parts of the country and freshwater lakes. At this season, numbers are taken in decoys and by other methods for the table, their flesh being highly prized.

Mr. Selby, who has had many opportunities of investigating the habits of the Teal in a state of nature, observes, that our indigenous broods " seldom quit the immediate neighbourhood of the places in which they were bred, as I have repeatedly observed them to haunt the same district from the time of their hatching till they separated, and paired on the approach of the following spring. The Teal breeds in the long rushy herbage, about the edges of lakes, or in the boggy parts of the upland moors. Its nest is formed of a large mass of decayed vegetable matter, with a lining of down and feathers, upon which the eggs rest: " they are eight or ten in number, and of a yellowish white. The young are at first covered with a dark-coloured down, which gradually gives way to a plumage differing little from that which is permanent in the adult female.

The plumage of the adult male, which is very beautiful, is as follows :

The top of the head, cheeks, and neck, of a deep chestnut; the throat black; from behind the eyes to the back of the neck passes a broad band of fine glossy green, margined by a pale yellowish border, into which the chestnut of the head and cheeks somewhat abruptly merges; the back, scapularies, and flanks rayed alternately with irregular zigzag bars of black and white; breast and under surface yellowish white, the former ornamented with round spots of black; wing-coverts brown; speculum glossy green, deepening at the sides into velvet black; quills brownish black; under tail-coverts buff, with a longitudinal band of black; bill black; irides brown; legs blackish brown. During the months of July and August, the male loses his finely contrasted plumage, and assumes that of the female, from which at this time he is not easily distinguished.

The female differs considerably, having the top of the head Sienna yellow, with dashes of deep brown; throat and cheeks dusky white spotted with brown; upper parts dull brown, each feather having a lighter border; under parts yellowish white; speculum green.

We have figured a male and female in the adult colouring, rather less than the natural size.

BIMACULATED TEAL.
Anas glocitans. *(Pall.)*
Querquedula glocitans. *(Vigors)*

Drawn from Nature & on stone by J.&E. Gould.

Printed by C. Hullmandel.

BIMACULATED TEAL.

Anas glocitans, *Linn.*

Querquedula glocitans, *Vigors.*

THE Bimaculated Teal is so named from the two large spots of brown on the face and neck: we believe, how-ever, that these brown markings vary in the depth of their colour at different seasons; at least we find such to be the case in an allied species from China, *Querquedula formosa*, of which examples are now living in the Gardens of the Zoological Society, and which possess marks of a similar character; these at opposite seasons are of a very different colour, changing from rich brown to light tawny grey. We are not aware of the existence of male and female examples of this very rare species in any collection, either public or private, except those in the Museum of the Zoological Society of London, to which they were presented with the rest of his fine collection by N. A. Vigors, Esq. These were taken in a decoy in the year 1812; and it is also recorded that a male was taken in a similar manner in 1771, as described by Pennant in his British Zoology. So rare does this Teal appear to be on the continent of Europe that we do not find it even alluded to by any writer except Pallas, who describes it as a native of the high northern regions of Siberia. In point of affinity this bird pos-sesses every characteristic feature of the true Teals, of which limited group it is the largest species that has come under our notice.

With regard to its habits, manners, and food, they are in all probability the same as in the other species of the genus. No account of them has yet been published; nor, indeed, is it to be expected that we shall easily acquire much information respecting the inhabitants of a portion of the globe so remote, and with which we have so little intercourse.

Crown of the head deep chestnut brown; sides of the head and neck rich green interrupted by two large blotches of brown, one situated near the base of the beak, the other on the side of the neck; chest rich chestnut regularly dotted with oval spots of black; the whole of the back and flanks light tawny grey, thickly pencilled with regular zigzag lines of black; shoulders greyish brown; quills blackish brown; speculum change-able green and blue, edged anteriorly with a narrow line of tawny yellow and posteriorly with a line of white; a row of coverts, which are internally edged with tawny yellow and externally with black, hang over the wing; rump and upper and under tail-coverts greenish black; two middle tail-feathers black, the rest pale brown, margined with white; a mark of buff separates the green under tail-coverts from the lower part of the belly, which is greyish white; bill olive brown, more yellow at the base; feet dark olive brown.

The female has the head and neck pale buff minutely spotted with small markings of black; the upper surface blackish brown, each feather having a margin of tawny brown; chest reddish brown, each feather being darker in the centre; shoulders of the wing as in the male; speculum green above with purple reflec-tions passing into black, and edged with white; quills and tail brown; the feathers of the latter edged with tawny white; under surface greyish white; legs more inclined to orange.

The Plate represents a male and female of the natural size.

GARGANY TEAL.
Anas querquedula. (Linn.)
Querquedula circia. (Steph.)

GARGANY TEAL.

Anas Querquedula, *Linn.*

Querquedula circia, *Steph.*

Le Canard sarcelle d'été.

THE feathers pendent from the back of this little Duck, together with its chaste and sober plumage, render it one of the most interesting and graceful species of its family. In point of affinity it has many characters in common with the genuine Teal, with which genus it has been by previous authors associated. It must be allowed, however, that it possesses some features in the style and markings of its plumage which are not in strict unison with the birds of that genus: this circumstance, and a slight deviation in the form of the bill, will in all probability hereafter lead to a further subdivision of the genus; in which case the Gargany, the Blue-winged Teal of America, and others, will form a minor group by themselves.

The range of the Gargany over the Old World is very considerable, being dispersed over the whole of Asia and Northern Africa, appearing to give preference to mountain districts, where it enjoys a temperature very similar to that of Europe, in every part of which it is abundantly distributed. It migrates annually to the British Islands during the months of April and May, and takes up its abode on our meres and large sheets of water, whence numbers are sent during the season to the London market, where they are esteemed as a great acquisition to the table, at a period when the Common Teal and most other edible species have retired to distant regions to breed. It is even questionable whether the Gargany that visits us at this period is not on its migratory route to more remote northern countries, such as Lapland, Russia, &c., where it may perform the task of incubation unmolested and in safety: we are strengthened in this opinion by the circumstance of their never being seen here during the autumn and winter; and even those individuals which visit us in the spring are extremely local in their habitat. Mr. Selby informs us that no instance is known of its occurring in the northern counties at any time.

Its food is strictly similar to that of the Teal and other ducks which are destitute of the power of diving, and consists of the tops and shoots of various aquatic plants, to which are added shelled snails, water insects, and their larvæ.

The nest is placed among herbage near the water; and the eggs, which are white, are from eight to ten in number.

The sexes, when adult, present a contrasted difference in their plumage. The young males during the first year, and the adult males during winter, are so like the female as to require an experienced eye to detect the difference.

The adult male in spring has the top of the head and back of the neck dark brown, a broad stripe of white extending over each eye; the cheeks and sides of the neck chestnut brown finely dotted with white; the lower part of the neck and chest light buff, each feather being marked with horseshoe-shaped lines of brown; the feathers of the back olive brown with lighter edges; the scapularies long, flowing, and of a green colour with a conspicuous stripe of white down the centre of each feather; the secondaries and shoulders light grey; the speculum green; the rump and tail brown, the former being spotted with darker brown; the belly white; the flanks transversely rayed with black and grey; and the feet and legs ash grey; bill blackish; irides brown.

The female has the top of the head, the back part of the neck, and the upper surface brown, the feathers having lighter edges; the throat white; the chest brown; the feathers edged with yellowish white; breast and under surface white tinged with buff; bill and feet blackish brown.

The Plate represents an adult male and female of the natural size.

PINTAIL DUCK.
Anas acuta; (Linn.).
Dafila caudacuta; (Leach).

Drawn from Life & on Stone by J & E. Gould.

Printed by C. Hullmandel.

Genus DAFILA, *Leach*.

Gen. Char.   *Beak* medial, its base unarmed, subcylindric, linear, its tip furnished with a very small hook : the *mandibles* with their edges lancinate, dentated.   *Nostrils* basal, suboval.   *Tail* elongated, acute, furnished with sixteen feathers.

PINTAIL DUCK.

Anas acuta, *Linn.*

Dafila caudacuta, *Leach.*

Canard à longue queue.

We have thought it best to give the genus and generic characters as established by Dr. Leach, leaving it to our readers to adopt it or not at their pleasure. The present bird will form the only known species belonging to it. If we except the Mallard, *Anas boschas*, Linn., the Pintail has a more extensive range than any other of its tribe. In Europe it is very generally distributed, as also in the northern portion of Africa, the whole of the Asiatic continent, and the northern and temperate regions of North America. On comparing specimens from all these different quarters of the globe, we can trace no distinguishing difference among them. It is one of the most graceful examples of its race ; although its colours are by no means remarkable for brilliancy or powerful contrast, yet its delicately penciled zigzag markings more than counterbalance its quiet and sober hues. Its form indicates it to be one of the true vegetable feeding ducks ; hence its flesh is peculiarly delicate and palatable. Although we state this bird to be a vegetable feeder generally, we believe, notwithstanding, that all the species of the Duck tribe subsist more or less on aquatic insects and molluscous animals. Its flight is rapid and vigorous, and its disposition in a state of nature is extremely wary and suspicious ; it is almost solely in the decoys of Lincolnshire, Cambridgeshire and Norfolk, that it is obtained in such abundance for the London market. It is said not to breed in the British Isles, and our own experience does not enable us to decide the point. It breeds nevertheless in considerable abundance in Holland, France and Germany, choosing morasses and vast reed beds for the site of its nest, which is placed on the ground concealed among the flags and luxuriant herbage near the water. The eggs are eight in number, of a greenish blue.

The male and female of this interesting species offer very considerable difference in their plumage. The male is characterized by the top of the head being variegated with black and brown ; the throat, cheeks, and upper part of the neck being brown with purple and violet reflections, a black band extending over the back of the neck, bordered on each side white ; the front of the neck and under parts of a pure white ; the back and sides barred with delicate zigzag lines of black and grey ; the speculum purple green, bordered above with rufous, and below with white ; the scapulars long and pointed, overhanging the quill-feathers, mostly of a deep velvet black with light grey edges ; the two middle tail-feathers of a greenish black, considerably elongated and tapering ; the beak blackish blue ; irides light brown ; feet dull reddish.

The female is known by her smaller size, and by the head and neck being of a light rufous dotted with small spots of black ; all the upper parts blackish brown, marked with regular crescent-shaped spots of reddish yellow ; the lower parts reddish yellow, spotted with light brown ; speculum reddish brown, bordered above with yellowish, below with white ; tail conical.

In the months of August and September the males resemble the females in the colour of the plumage.

We have figured a male and female two thirds of the natural size.

GADWALL.
Anas strepera; *(Linn.)*
Chauliodes strepera; *(Swains.)*

Genus CHAULIODES.

GEN. CHAR. *Bill* as short as the head, depressed throughout its length, as broad as high at the base, rather narrowing towards the tip, which has a small *dertrum* or *nail*; both mandibles laminated, the laminæ of the upper one projecting beyond the margins of the bill. *Nostrils* lateral, near the base of the bill, oval and pervious. *Wings* long and acuminate. *Tail* wedge-shaped. *Feet* with four toes, three before and one behind; the hind toe small and free.

GADWALL.

Anas strepera, *Linn.*

Chauliodes strepera, *Swains.*

Le Canard chipeau.

THOUGH the colours of this elegant Duck are more sober than those of most of the family, it yields to none in the tasteful disposition of its markings, and to few in the excellence of its flesh for the table. The European species to which it is nearest allied is the Common Widgeon (*Anas Penelope*), and we can scarcely see the necessity of creating a new genus for its reception from which the Widgeon is excluded: we have, however, given Mr. Swainson's generic characters, leaving it to the option of our readers whether to accept or reject them.

Although the Gadwall does not visit us in great numbers, it is tolerably common during the months of spring. The low marshes and fenny districts, Holland, and the whole of the northern portions of Europe, are the situations in which it most abounds. In its habits and manners it closely resembles the Widgeon, with which it is often seen associated. We have received specimens from the Himalayan mountains which are identical with our European species.

The deficiency of brilliant colours in the male renders him but little more ornamented in his plumage than the female, the external difference between them being less than is usually met with; but the bird of the first year presents a considerable difference, as our Plate illustrates.

Like the common Wild Duck, this bird breeds in reed beds and similar places, laying eight or nine eggs, of a pale green.

The adult male has the head and the upper part of the neck dull brown, thickly marked with dirty white; the back, scapulars and sides ornamented with narrow zigzag lines of black and white; the middle wing-coverts chestnut, with a dash of brown in the centre; the rump and under tail-coverts blueish black; the shoulders chestnut, succeeded by blueish black, and a white speculum; the feathers of the chest scale-like, having a dusky black centre, with crescent-shaped edges; the abdomen white; the beak black, and the tarsi orange.

The young bird of the year is of a uniform rusty brown above, each feather having a central mark of dusky black, the under surface being white.

We have figured an adult male, and a bird of the first year, three fourths of the natural size.

RED-HEADED POCHARD.
Anas ferina; *(Linn.)*
Fuligula ferina; *(Steph.)*

RED-HEADED POCHARD.

Fuligula ferina, *Steph.*

Le Canard Milouin.

This fine species may be said to represent in Europe the Canvas-backed Duck of America, so famed for its rich and juicy flesh ; and although the flesh of the Pochard is superior to that of all the other European diving ducks, still it must, we are told, yield the palm in this respect to its Western ally.

The Red-headed Pochard is an article of considerable traffic in the London markets, where it is known by the name of the Dunbird. So vast is the quantity taken during the year, that, were our information not received from an undoubted source, we should have hesitated in stating the amount ; but we are positively assured that no less than fourteen thousand four hundred have been captured in one decoy, the sale of which produced twelve hundred pounds.

Although this species is frequently taken in the usual decoys, still, we are informed by Montagu, the method commonly practised was something similar to that of taking woodcocks. Poles were erected at the avenues to the decoy, and after a great number of these birds had collected on the pool, a net was erected by pulleys to the poles, beneath which a deep pit had previously been dug ; and as these birds, like the wood-cocks, go to feed just as it is dark, and are said always to rise against the wind, a whole flock has been taken together in this manner ; for when once they strike against the net, they never attempt to return, but flutter down till they are received into the pit, from whence they cannot rise.

The Red-headed Pochard is very widely dispersed, being common over the whole of Europe, Asia, and a portion of Africa. It is said to breed in the marshes, and to lay about twelve white eggs. Its food consists of aquatic vegetables, mollusca and other animals, obtained by diving to the bottom, which it does with a facility only equalled by its vigorous flight. Being entirely aquatic in its habits, it not unfrequently takes up its abode on the open sea, where it obtains a plentiful supply of bivalves and other shells, of which it appears fond, but which kind of food generally gives a fishy and unpleasant flavour to its flesh.

The male has the head and neck chestnut brown ; the breast and rump black ; the back, scapulars, wing-coverts, thighs, and flanks greyish white, beautifully pencilled with zigzag lines of black ; the quills and tail grey ; the bill blackish grey with the tip and base black ; and the tarsi and toes bluish grey.

The female has the head and neck of a dark reddish brown ; the under surface dusky white ; and the back like that of the male, except that the whole of the black markings are darker and more obscure.

The Plate represents a male and female rather less than the natural size.

WHITE-EYED OR CASTANEOUS DUCK.

Anas leucophthalmos. *(Bechst.)*

Fuligula ———. *(Steph.)*

Drawn from Life & on Stone by J.& E. Gould.

Printed by C. Hullmandel.

WHITE-EYED OR CASTANEOUS DUCK.

Anas leucophthalmos, *Bechst.*

Fuligula leucophthalmos, *Steph.*

Le Canard à iris blanc.

THIS interesting little Duck has been several times killed in England: its occurrence, however, in the British Isles, which is generally during winter, must be considered accidental rather than as that of a regular visitor. It is more abundant in France, Holland and Germany, in the latter of which countries it appears to be a periodical bird of passage. We have received it in abundance from India, especially from the elevated range of the Himalaya, and it appears from the accounts of Buffon and Sonnini to be equally common in the North of Africa.

It is much less in size than the Pochard, to which it bears a close affinity. It is the Ferruginous Duck of Pennant and Montagu, but not of Bewick, his figure and description applying to the *Anas rutila*. Its habits and manners are strictly analogous to those of the Pochard, being an expert diver and living upon aquatic insects, water-plants, small shell-fish, &c. Like most of the pointed-winged Ducks, its power of flight is very considerable.

M. Temminck informs us that it constructs its nest among reeds by the sides of large rivers and morasses; that the eggs are eight or ten in number, of a white colour tinged with greenish.

The sexual diversity of plumage is not so considerable in this species as in many others of the Duck tribe.

The male has the head, neck, breast and sides of a rich bright reddish chestnut; a slight collar of deep brown encircles the neck; beneath the lower mandible there is a small triangular spot of pure white; the back and wings are of a blackish brown with purple reflections, covered with small reddish dots; speculum white, banded with a line of black; under parts pure white; beak blueish black; irides clear pearl white; tarsi blueish ash colour; webs black.

The female has the head, neck, breast and sides of a dull brown, inclining to chestnut; the under parts of an obscure brown, each feather having a light brown termination; and is destitute of the dark brown collar round the neck.

The young of the year have the top of the head blackish brown, all the feathers of the upper parts edged with reddish brown, and the white of the under part clouded with a lighter tinge of the same colour.

The trachea of the male is very narrow at the top, and also just before its termination in the inferior larynx, but of double the diameter at the middle: the inferior larynx is formed of an osseous wall on the right side, and on the left presents a series of bony ramifications supporting an external membrane.

We have figured a male and female two thirds of their natural size.

RED-CRESTED DUCK.
Anas rufina (Pallas)
Fuligula rufina (Steph.)

RED-CRESTED DUCK.

Anas rufina, *Pall.*

Fuligula rufina, *Steph.*

Le Canard siffleur huppé.

THE very fine Duck which we have illustrated in the accompanying Plate is as yet but little known as having claim to a place in the Fauna of Great Britain; but the frequent occurrence of both sexes in various parts of the British Islands sufficiently establishes it as a native species, or at least as much so as many others that occasionally migrate to this country.

English examples of this beautiful species form a part of the collections of the Hon. W. T. T. Fiennes and Mr. Yarrell. The former gentleman possesses a fine female, killed out of a flock of eighteen, on the Thames, near his own estate at Erith in Kent, and to whose kindness we are indebted for the loan of the specimen from which our figure was taken.

The *Anas rufina* is confined to the old continent, where its range is very extensive, as is proved by our having received it in collections from the Himalaya mountains, and observed it in the collection of Col. Sykes from the Dukhun, in which localities it is a bird of no rarity; and it also occurs nearly as plentifully in the eastern portions of Europe, particularly throughout Hungary, Austria, and Turkey. M. Temminck states that it is a periodical visitor to the shores of the Caspian Sea, but at the same time observes that it never visits the open ocean: from these countries it is more or less distributed throughout the whole of the central portions of Europe. Little is known of the habits of this very interesting species: its form, however, shows it to belong to the true diving Ducks; hence we may reasonably conclude that its food consists principally of small shell fish and molluscous animals, with vegetables and the fry of fishes.

An attentive examination of this bird will lead, we think, to the conviction, that it offers many points of affinity to the species of the genus *Mergus*. We need only instance the narrow and compressed form of the bill towards its extremity, with deeply serrated edges, the disposition of some of its markings, and the silky texture of the feathers of the head, in corroboration of this fact. The trachea of the male, also, according to M. Temminck's description, is not unlike that of the *Mergus merganser*, being large immediately below the upper larynx, becoming suddenly very narrow, and then a second enlargement of the tube, terminating in very narrow rings. The inferior larynx is formed of two dilatations: that on the left, which is the largest and most elevated, is formed of osseous ramifications covered by a fine membrane.

The male has the head ornamented with a crest of silky feathers, which, with the rest of the head and the front of the upper part of the neck, is of a delicate chestnut tinged with vinous; the back and lower part of the neck, the chest, and under surface, are brownish black; the back is pale cinereous brown, with a large spot of white above the origin of each wing; the shoulders, the speculum, the base of the quills, and the flanks, are white; the rump and upper tail-coverts black with green reflections; beak red; nail white; tarsi and toes red with black interdigital membranes.

The female wants the fine crest of the male; the top of the head and occiput are dark brown; cheeks, throat, and sides of the neck, cinereous; the whole of the upper surface cinereous brown, with the exception of the shoulders, which are white, and the operculum, which is dull white terminating in brown; breast and flanks yellowish brown; under surface cinereous; beak, tarsi and toes, reddish brown.

We have figured a male and female three fourths of the natural size.

TUFTED DUCK.
Anas cristata. *(Ray.)*
Fuligula cristata. *(Steph.)*

TUFTED DUCK.

Anas cristata, *Ray*.

Fuligula cristata, *Steph*.

Le Canard Morillon.

THIS elegant little Duck is one of the regular winter visitors of our island, where it arrives in autumn and distributes itself very generally over lakes, meres, large ponds, armlets of the sea, and similar situations, in which temporary residences it is most frequently seen in pairs, repeatedly diving in search of food, which is obtained exclusively at the bottom of the water, and consists, for the most part, of various freshwater shell-fish, crustacea, worms, and mollusca; to this food it occasionally adds aquatic vegetables. Although generally observed inland in pairs, it is not unfrequently to be seen in considerable flocks enlivening the shores of the sea, particularly those of a rocky nature. In the power of diving, the Tufted Duck possesses the same facility as the rest of the Lobefooted section to which it belongs, being extremely quick in all its motions, and on this account difficult to be shot. On the approach of spring it retires northward to breed, and like many of its congeners makes the morasses and the unfrequented regions of the arctic circle an asylum in which to rear its young. The range of its migrations southward is very considerable: we ourselves have received it from the southern portions of Europe, and also from those parts of the Mediterranean which are near the Asiatic confines; we have also seen it from the Black Sea, and from every intermediate country as far as the high northern latitudes. It abounds in many parts of Northern India, especially the high lands. The collection from the Himalaya mountains which afforded the materials for our 'Century of Birds,' contained several specimens. Although its stout and rounded figure does not contribute much to its elegance, its plainness is relieved by the graceful pendent crest which flows from the head and occiput.

The Tufted Duck is brought to the London market in considerable numbers during the winter; and although often seen at table, its flesh is not, we believe, of the most delicate quality.

The sexes may be distinguished by the lesser comparative brilliancy of the colours of the female, and by her having the rudiments only of the flowing crest, the colour of which, as well as of the breast, is brown, with faint traces of the violet gloss so conspicuous in the male, whose plumage may be thus described:

The top of the head and the long pendent silky crest rich violet black; neck and chest greenish black; upper part of the plumage glossy brownish black, with very minute grey specks scattered over the scapularies; speculum of the wings and under surface white; bill lead colour; nail black; legs and toes brownish black.

The young of the year are devoid of the pendent crest; the whole of the plumage is of a more obscure tint; and the feathers of the upper surface are edged with brown.

The plate represents a male and female rather less than the natural size.

SCAUP POCHARD.
Fuligula marila. (Steph.)
Anas marila. (Linn.)

Drawn from Nature & on Stone by J.&E.Gould.

Printed by Hullmandel

SCAUP POCHARD.

Fuligula marila, *Steph.*

Anas marila, *Linn.*

Le Canard Milouinan.

THE native residence of the Scaup Duck during the summer season is within the regions of the Arctic circle: it is in these high latitudes that it breeds and rears its young. On the approach of winter it is driven southward, and appears in vast flocks in many parts of the European shores, and especially on those of Great Britain, Holland, France, &c. In its habits it appears to be essentially marine, consequently it is seldom seen even on the larger of our inland lakes, but our bays and the mouths of rivers are frequented by it during the winter in considerable abundance: from these its temperate places of refuge it retires early in spring to the latitudes from whence it came; in fact, so universally is this the case that we do not know of a single instance of its breeding in our island. We are not acquainted with the details of its nidification, but in this respect it doubtless agrees with the diving ducks in general.

Its food principally consists of univalves, bivalves, mollusca, marine plants, &c., which it obtains by diving, a power it possesses in a very great degree.

As an article of food the Scaup Duck is inferior to most of the genus to which it is assigned; still it is not unfrequently sold for the purposes of the table.

" It makes a hoarse noise, and has a singular habit of tossing its head and opening the bill, which is more particularly observable in spring, while it is swimming and sporting about on the water."

The sexes differ so much in plumage that the female has been described as a distinct species, under the name of *Anas frœnata.*

The male has the head and upper part of the neck black, with reflections of rich glossy green; lower part of the neck, breast, and rump deep black; mantle and scapulars greyish white with distant zigzag fine lines of black; lesser wing-coverts black with transverse zigzag lines of white; secondaries white with black tips, which form a bar across the wings; under surface white, the belly rayed with lines of blackish grey; bill greyish blue with the nail black; irides light yellow; legs and toes bluish grey, the joints and webs darker.

The female has a broad band of white round the base of the bill; the remainder of the head and upper part of the neck deep brown glossed with green, lower part of the neck and breast having the basal part of the feathers brown, the tips of the former deeply margined with yellowish brown, and those of the breast with white; under surface white; flanks brown, marbled with zigzag lines of white and darker brown; mantle and scapulars brownish black finely mottled with white; tertials black, tinged with green; quills and tail glossy blackish brown; bill deep grey; nails black.

The young males resemble the old female till after the first moult. The colours of the young females are less distinct, and the black and white lines on the back are scarcely perceptible.

We have figured a male and female rather less than the natural size.

WESTERN DUCK.
Anas Dispar. (Gmel.)
Fuligula Dispar. (Steph.)

WESTERN DUCK.

Fuligula Dispar, *Steph.*

From the circumstance of an example of this very rare Duck having been obtained in February 1830 at Caisted near Yarmouth in Norfolk, about which period another specimen was also killed in Denmark, we have much pleasure in figuring it as an interesting addition to our native fauna. The drawing from which our figure of the first-mentioned specimen (now in the Norwich Museum) is taken, was presented by Joseph Clarke, Esq., to the Saffron Walden Natural History Society, in whose Museum it is deposited, and to whom we are indebted for the loan of the drawing.

The above is the only instance of its capture in Great Britain, and it appears to be equally rare on the Continent. It is a native of Northern Asia, Siberia, Kamtschatka, and the north-western coasts of America : it is said to breed in high and precipitous rocks, and to fly in extensive flocks.

We have followed Mr. Selby and others in placing it provisionally in the genus *Fuligula*; for like that gentleman, not having seen the bird itself, we are unable to decide upon its true station among the *Anatidæ*; but from its general contour, the disposition of its colouring, and the curved form of the tertials, we should conceive that it more properly belongs to the genus *Somateria*.

Space between the bill and the eye green, and on the back of the head a patch of the same colour, forming a short crest; throat and an irregular spot behind the eye black; the remainder of the head and neck white; throat surrounded by a band of black glossed with green; back, rump, vent, and under tail-coverts black; primaries and tail brownish black; lower part of the neck, part of the scapulars, and lesser wing-coverts white; breast and all the under surface rufous, becoming darker as it approaches the vent; tertials long and curved; the shaft and the narrow inner web white; the outer web broad and deep bluish black; irides pale brown; bill and legs blackish grey.

Our figure is of the natural size.

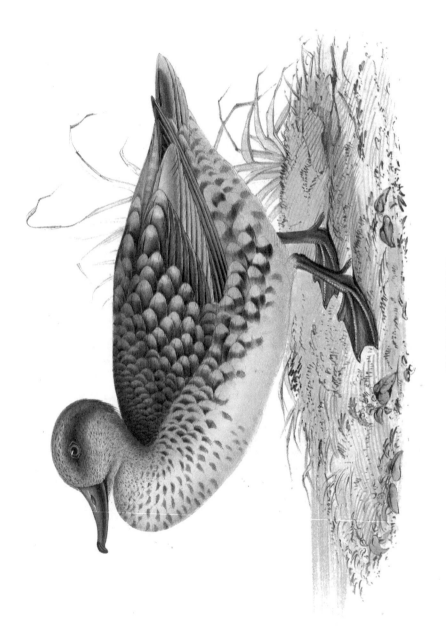

MARBLED DUCK.
Anas (Fuligula?) marmorata, (*Temm.*)

Drawn from Nature & on Stone by E.F. Gould.

Printed by C. Hullmandel

MARBLED DUCK.

Anas (Fuligula) marmorata, *Temm.*

Le Canard marbré.

FOR the specimen of this bird from which the accompanying figure was taken we are indebted to the kind friendship of M. Temminck of Leyden, whose valuable works on Natural History are duly appreciated. We cannot pass over this opportunity of acknowledging the great liberality of this justly celebrated naturalist, who has taken so much interest in the present work that he has promoted it by every means in his power, not only by his individual patronage, but by confiding to us for our illustration many rare species peculiar to remote districts of Europe, among which is the present species. Of its habits and manners we have no detailed account. M. Temminck, however, assured us that the sexes offer no difference in the colours of the plumage, a circumstance which we should not have expected, judging from the affinity it bears to the *Anas rufina*, which, although at present comprehended in the genus *Fuligula*, may be said to possess characters which claim for it a distinct generic station. Such genera, however, are of a subordinate character and value, and although of real utility to the professed ornithologist, are less likely to interest the general reader.

The only information we can communicate respecting the present bird is, that it inhabits the southern districts of Europe, particularly Sardinia and the Asiatic borders.

The crown of the head, back of the neck, the whole of the upper surface, flanks, and tail, of dull ashy brown, each feather being tipped with dirty white; outer webs of the quill-feathers greyish brown, tips of the inner webs the same colour as the upper surface; secondaries pale brown; cheeks and throat marked in the same manner as the upper surface, but much lighter; breast and the whole of the under surface dirty white, each feather being barred near its extremity with dull ashy brown, which, together with the white tips of the feathers on the upper surface, gives it somewhat the appearance of marble, whence its name; bill and feet dark brown.

The Plate represents an adult bird rather less than the natural size.

EIDER DUCK.
Anas mollisima. *(Linn.)*
Somateria mollisima. *(Leach.)*

Genus SOMATERIA, *Leach.*

GEN. CHAR. *Beak* swollen at the base, elevated, extending up the forehead, and divided by a triangular projection of feathers; towards the tip narrow and blunt. *Nostrils* small, placed in the middle of the beak.

EIDER DUCK.

Anas mollissima, *Linn.*

Somateria mollissima, *Leach.*

Le Canard Eider.

THE Eider Duck in its wild state is one of those birds which confer important services upon the human race. Its soft and exquisite down is an extensive article of commerce; and so great is the demand for it, that the inhabitants of the northern islands of Great Britain, together with those of Lapland, Iceland and Greenland, use every means to encourage the Eider to breed on their shores, in order that they may obtain from its nest this valuable material. It is scarcely ever found to incubate on the main land, but chooses the small islands scattered along the coast; and of this disposition the inhabitants take advantage by insulating small portions of ground, which enables these birds to perform their work unmolested by cattle, dogs, foxes or other wild animals, which appear to cause them great annoyance. The female is very assiduous in her work of nidification. The nest is placed on the ground, and constructed of soft down which she plucks from her own breast and under surface. This light and elastic material is so ingeniously disposed as to form an elevated rim round her body while sitting, and to fall over the eggs the moment she leaves the nest; nor is the quantity of the material less remarkable. So absorbed are the birds in this important duty, that they appear to lose all sense of danger, and may sometimes be approached and even taken off the eggs without attempting to escape. As soon as the nest is constructed, the plunder of the down commences. The first portion is taken away, and a fresh quantity is again afforded by the female bird, and again taken, till she is unable to furnish a further supply, when, as it is asserted, the male makes up the deficiency.

The task of incubation appears to devolve principally upon the female, the male being seldom seen near the nest by day; and so well does her sober colour harmonize with that of the surrounding earth, that she is a much less conspicuous object than the male would be, adorned as he is with a light and showy plumage; nevertheless as evening approaches he may be seen returning from the sea to his mate, whom it is not improbable he relieves during some portion of the night. The eggs are five in number, of a uniform olive-green. As soon as the young are hatched, they are taken by the parents to the sea, in which they find at once both food and security.

The Eider Duck is generally dispersed along the northern shores of Europe, and in the same latitudes of America, being more abundant within the arctic circle; they are often seen, associated in numerous flocks, diving in search of their food, which consists of shell-fish (particularly the common mussel), crustacea, insects, the ova of fishes, and marine vegetables.

It does not appear to be migratory, although it is not improbable that it is often driven southward by stress of weather.

On each side of the head and above the eyes there extends a very large band of black velvet-like feathers, the extremities of which unite over the forehead; the occiput and back part of the cheeks sea-green; the lower part of the neck, back, scapulars and lesser wing-coverts white with a tinge of yellow; the breast of a light buff; under parts and rump of a deep black; beak and feet olive. The adult female has all the plumage of a brownish red, barred transversely with black.

We have figured a male and a female, three fourths of the natural size.

Printed by C. Hullmandel

KING DUCK.
Anas spectabilis, (Linn.)
Somateria spectabilis, (Leach)

Drawn from Nature & on Stone by J & E Gould.

KING DUCK.

Anas spectabilis, *Linn.*

Somateria spectabilis, *Leach.*

Le Canard à tête grise.

This magnificent species has, with the Eider Duck, to which it bears a strong resemblance, been very properly formed into a distinct genus by Dr. Leach; which, although it contains only these two species, as far as known, is yet marked by well-defined characters. The habits, manners, and localities of the King Duck closely resemble those of the Eider, excepting that it seldom visits, like the latter, our more temperate latitudes, but confines itself more exclusively to the seas of the Arctic circle. Its claims to the rank of a British species rest upon its occasional capture upon our coast; it is, however, stated in Dr. Latham's "General History of Birds," that Mr. Bullock discovered it breeding in Papawestra, one of the Orkney Islands, in the latter end of June; but such an occurrence we suspect could only have been accidental, as we are not aware of a similar circumstance being recorded. It is not uncommon in Norway, the coasts of the Baltic, the arctic shores of Siberia, and it even extends to Kamschatka. It is very abundant in Greenland, where its flesh is eaten by the natives, and its skin sewn together to form warm garments. It is also dispersed in equal abundance along the same latitudes of the American coasts. The process of incubation is strictly the same as that of the Eider, and the nest undergoes the same system of plunder. Its eggs are rather less in size, and of an olive colour.

Mr. Sabine, in his history of the Birds of Greenland, informs us, that the male is four years in acquiring its perfect plumage. The sexual differences in this species present the same features as are found to occur in the Eider; the female being of a dull sober brown, while the male is as remarkable for a gaudy and strongly contrasted livery.

The beak of the male bird differs from that of the male Eider in having two lateral cartilaginous projections which rise from the base, inclosing the forehead nearly as far as the eye; the colour of its crest, as also that of the beak and legs, is of a beautiful rich vermilion; a narrow line of black velvet-like feathers forms the outline of the upper mandible; beneath the throat are arrow-shaped lines of the same colour pointed towards the base of the beak; the top of the head and the occiput are of a beautiful blueish grey; cheeks white, with a delicate tint of fine sea-green; neck and upper part of the back white, gradually assuming a delicate salmon colour at the breast; the other parts of the plumage, both above and below, deep blackish brown, with the exception of a white spot in the centre of the wing, and another of the same colour behind the thigh; the secondaries are sickle-shaped, and bend gracefully over the quill-feathers.

So closely does the female of this species resemble the female Eider, that it is almost impossible to distinguish them, being of a uniform rusty brown with irregular bars and arrow-shaped markings of black.

We have figured a male and female three fourths of their natural size.

SURF SCOTER.
Anas perspicillata, (Linn.)
Oidemia perspicillata, (Flem.)

Genus OIDEMIA, *Flem.*

GEN. CHAR. *Bill* swollen or tuberculated at the base, large, elevated, and strong; the tip much depressed and flattened, terminated by a large flat dertrum or nail, which has its extremity rounded and slightly deflected; mandibles laminated, with the plates broad, strong, and widely set. *Nostrils* lateral, elevated, oval, placed near the middle of the bill. *Wings* of mean length, concave, acute. *Tail* short, graduated, acute. *Legs* far behind the centre of gravity. *Tarsi* short. *Feet* large; of four toes, three before and one behind; outer toe as long as the middle one, and much longer than the tarsus; hind toe with a large lobated membrane.

SURF SCOTER.

Anas perspicillata, *Linn.*

Oidemia perspicillata, *Flem.*

Le Canard Marchand.

THIS curious Duck should rather be considered as an American species than as strictly indigenous to the European Continent; it has, however, frequently occurred in the northern seas of this portion of the globe, and occasionally as far south as the Orkneys and other Scottish islands: we have ourselves received a specimen (a female) killed in the Firth of Forth. In its general form, economy, and habits, it is intimately allied to both the Velvet and Scoter Ducks, and the three species have been with good reason separated by Dr. Fleming into a distinct genus. No one who has attentively investigated the great family of the *Anatidæ* can have failed to remark into how many distinct groups or genera even the European examples naturally arrange themselves, each group being characterized by its diversity of form, habits, and manners. Of these genera, one of the best defined as well as most conspicuous is that designated *Oidemia*. The species of this genus are strictly oceanic, and are expressly adapted for obtaining their food far from shore, being provided with an entirely water-proof plumage, and endowed with most extraordinary powers of swimming and diving. Unlike the true Ducks, they seldom visit the inland waters, or feed upon terrestrial mollusca or vegetables, but keep out at sea, and diving to a very great depth, procure bivalves, mollusca, and submarine vegetables: they appear to be particularly partial to the common mussel, which we have taken from their throats and stomachs entire.

It is for the purpose of grinding down this shelly food that the gizzard is not only extremely thick and muscular, but is also lined with a dense coriaceous cuticle capable of grinding to pulp the hard bodies subjected to its action. The arctic regions of America appear to be the true habitat of the present species, particularly about Hudson's Bay and Baffin's Bay.

Little is known respecting its nidification, but it is said to form its nest near the shore, of grasses lined with down; and that the eggs are white, and eight or ten in number.

The wings are short, convex, and pointed, and although they afford the bird tolerable powers of flight, they are equally adapted for an organ of progression under water, an element to which, rather than to the air, it frequently trusts for safety.

The uniform black colouring which characterizes the plumage of the present group is relieved in all the species by a beak exceedingly rich in colour and ornamental in its markings, and in none of the species is this peculiarity more conspicuous than in the Surf Scoter; this feature, however, is found in the males alone, the females of the three species, which, we may remark, very closely resemble each other, having the beak plain, and nearly uniform in colour.

The adult plumage, which presents no difference in summer and winter, may be thus described:

The male has the bill scarlet and yellowish white, with a large black mark on each side of the swollen basal portion; the whole of the plumage glossy black, with the exception of a patch of white on the top of the head and another on the occiput; the irides greyish white; legs and toes red; interdigital membrane black.

The female differs from the male in having the whole of the plumage dull brown, which is lightest about the face, cheeks, and under surface; the beak dark olive; feet greyish brown.

The Plate represents a male and female, rather more than three fourths of the natural size.

VELVET SCOTER.
Oidemia fusca. (*Flem.*)

VELVET SCOTER.

Oidemia fusca, *Flem*.

La grande ou double Macreuse.

This is the largest species of the genus *Oidemia*, and may be readily distinguished from both its congeners (*Oid. perspicillata* and *Oid. nigra,*) by the snow-white bar across the wing and the patch of white situated beneath the eye. It also differs very considerably in the conformation of its bill, in which, however, it approximates most nearly to *Oid. perspicillata ;* for although it is much more dilated, still it presents traces of the swollen tubercle, but to a less extent than in either of the other species.

The northern regions of the globe constitute the true habitat of all the members of the group ; the present species retires within the Arctic Circle during the summer, and regularly migrates to the southern seas on the approach of severe weather. It is found in considerable abundance on various parts of our northern coasts, where it feeds upon muscles, mollusca, &c., which it always obtains by diving. It is very abundant in the Arctic regions of Asia, and is reported to breed upon the banks of the larger rivers in Denmark, Russia, Kamtschatka and Siberia : it is also found in America. It generally goes far inland for the purpose of nidification, and constructs its nest of grass, lined with down, in which it deposits from eight to ten white eggs.

As in the other members of the genus the sexes differ very considerably from each other, but as the female very closely resembles the female of the Surf Scoter we have not deemed it necessary to give a figure of it.

The male has the whole of the plumage of a deep velvety black, with the exception of a patch of white beneath the eye, and the secondaries, which are pure white and form a band across the wing ; upper mandible black at the base, the remainder red, edged with black ; under mandible pale yellowish white, edged with black ; legs red on the outside, and deep yellow sprinkled with black on the inner ; irides white.

The female has the whole of the plumage of a dull blackish brown, which is much lighter and rayed with greyish on the under surface ; between the bill and the eye, and on the ear-coverts, a spot of white ; beak blackish ash colour ; tarsi and feet dull red ; irides brown.

We have figured a male of the natural size.

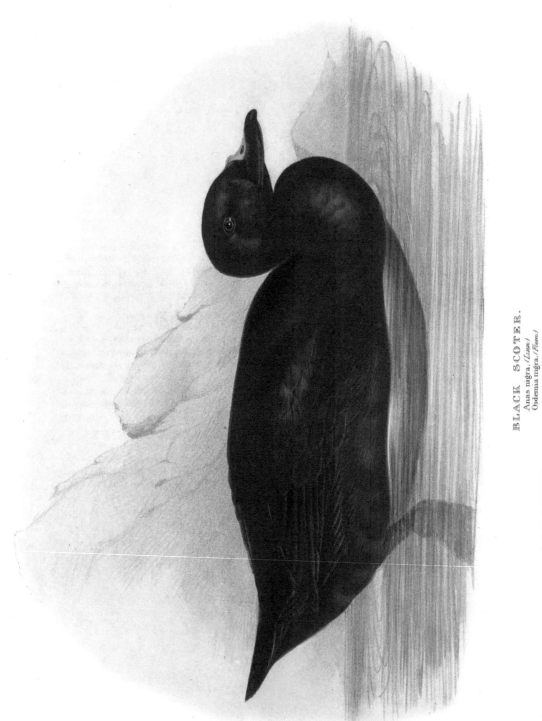

BLACK SCOTER.
Anas nigra. (Linn.)
Oidemia nigra. (Flem.)

BLACK SCOTER.

Anas nigra, *Linn.*

Oidemia nigra, *Flem.*

Le Canard macreuse.

No one of this truly oceanic group of Ducks is more familiar to our readers than the Black Scoter, visiting as it does, during its spring and autumnal migrations, the seas which immediately encircle our island; in fact, we can seldom at these seasons cross the channel between England and the Continent without observing it; numerous flocks of them winging their way from one shoal or fishing-place to another, or diving in pursuit of their prey. They subsist almost entirely on bivalves, such as the common mussel, &c., and they especially abound where large beds of these shell-fish afford them an unfailing supply of favourite diet, their close ad-pressed plumage and great power of diving admirably fitting them for their destined mode of life. Although so plentiful on our shores, and on those of the Continent, especially Holland, it does not appear that it ever breeds in our latitudes, but retires for that purpose to the seas, lakes, and morasses of the arctic circle, whence it is annually driven southwards as winter locks up these waters and precludes the possibility of its obtaining its natural food. Of its nidification we have no positive information, as is also the case with most of those birds that resort to the higher regions to breed.

Unlike most of the *Anatidæ*, the Black Scoter and its allies undergo no periodical change in their plumage; neither is there so great a dissimilarity between the opposite sexes as there is in most others of this family, the bright colouring of the bill in the male and his more richly coloured plumage being the chief points of difference.

The male has the whole of the plumage of a rich velvet black; the beak black, with the exception of the nostrils, which are bright orange, and the spherical protuberance at the base, which is banded with yellow; irides brown; naked circle round the eye red; tarsus and toes brownish ash colour; webs blue.

The female is characterized by a plumage of dull blackish brown; the bill black, tinged with olive, and wanting the basal protuberance; the sides of the face, throat, and under surface lighter in colour than the upper.

The Plate represents an adult male of the natural size.

GOLDEN EYE.
Anas clangula. (Linn.)
Clangula vulgaris. (Leach.)

Drawn from Life and on stone by J.&E. Gould.

Printed by C. Hullmandel

Genus CLANGULA.

GEN. CHAR. *Beak* shorter than the head; its base simple, narrow, nearly straight towards the tip; the upper *mandible* furnished with a small hook. *Nostrils* oval, basal. *Tail* with its feathers acuminated or blunt, not elongated.

GOLDEN EYE.

Anas Clangula, *Linn.*

Clangula vulgaris, *Leach.*

Le Garrot.

OF all the diving Ducks the Golden Eye displays the most address in the water, the greatest rapidity in plunging, united to the power of long continuance beneath its surface; and as its food is to be sought for only at the bottom of the deep, we see in these qualifications one of those instances which Nature ever presents of the adaptation of the means to the end.

This interesting bird is a winter visitor, arriving on our coasts and those of the neighbouring continent at the latter end of autumn, and retiring northwards to breed, as the milder weather approaches taking up its summer residence in Norway, Sweden, and the arctic portion of the American continent; the female, it is said, lays from ten to fourteen eggs of a pure white, on the borders of lakes and inlets of the sea.

The immature males of the Golden Eye, as well as the adult females, have been characterized as a distinct species under the name of *Morillon* or *Anas Glaucion,*—a mistake lately rectified. But though the young males and females present no dissimilarity of colour, the anatomical structure of the trachea, which exhibits the same peculiarities in the males of every age, and which may be felt externally, is an infallible criterion. The singularity of this organ in the present species consists of a labyrinth, very irregular in figure and almost entirely osseous, from which the bronchial tubes proceed;—a short distance above this, the trachea itself enlarges very considerably; the dilatation is of an oval figure, capable of extension and contraction, and formed of rings placed in an oblique direction. In the males of all the Ducks the trachea presents a peculiarity of structure, which differs in every species: this feature therefore affords a clue to specific distinction, and may be taken as the best test for identifying an immature or doubtful species. The use for which Nature has designed this peculiarity of structure has not been clearly ascertained, but it is most probably connected with the tone or modulation of the voice.

In its mode of living; in the disproportion that exists between the sexes; in the dark green tufted head of the male,—a colouring exchanged for brown in the female and young,—there would seem to be indicated we think, a striking analogy between the Golden Eye and the less typical Mergansers. This resemblance is the more apparent if we take the Smew for our comparison. We there find a shorter beak, a more rounded contour of body, and a less brilliant colour pervading the feet and tarsi, than in the others of its genus: to this we may add its food, which consists more exclusively of molluscous animals and crustacea. The food of the Golden Eye is the same, for which its strong beak, tapering from a thick base, is well constructed. The colour of its plumage also partakes of the same character as the Smew, exhibiting a contrast of black and white in the male, and dark grey and white in the female.

The bill of the Golden Eye is black; irides fine golden yellow; neck glossy greenish black, with the exception of a large white spot at the base of the bill; back and tail black; a band of white crosses the wings, leaving the shoulders and quills black; legs dull orange; webs darker. The female is considerably less than the male, and has the bill yellow towards the point; head and upper part of the neck rusty brown, below which is a ring of greyish white; breast mottled with grey, upper parts dark cinereous; tail and under parts as in the male.

The weight is nearly two pounds, the length seventeen or eighteen inches.

Our Plate represents a male and female, faithfully figured, two thirds of their natural size.

BARROW'S DUCK.
Clangula Barrovii (Swains.&Rich.)

BARROW'S DUCK.

Clangula Barrovii, *Swains.* and *Rich.*

A FINE male of this rare and beautiful species having been shot in Iceland by T. C. Atkinson, Esq. of New-castle-upon-Tyne, when on a visit to that country, about two years since, we have much pleasure in figuring it as an occasional inhabitant of the European portion of the globe.   Mr. Atkinson's specimen is now deposited in the Museum of the Newcastle Natural History Society.   Although very nearly allied to the Golden Eye (*Clangula vulgaris*), it possesses, nevertheless, many characters by which it may be distinguished from that species.   For our first knowledge of the *Clangula Barrovii* we are indebted to that highly interesting work the " Fauna Boreali-Americana" of Messrs. Swainson and Richardson ; and as the observations of the latter gentleman were taken on the spot, we cannot do better than extract what he has recorded in the work above mentioned.

" Notwithstanding," says Dr. Richardson, " the general similarity in the form and markings of this bird and the Common Golden Eye, the difference in their bills evidently points them out to be distinct species. Exclusive of other specific characters," the Barrow's Duck " is distinguished by the purer colour of its dorsal plumage, and the smaller portion of white on its wings and scapulars.   Its long flank feathers are also much more broadly bordered all round with black.   The bases of the greater coverts in the Golden Eye are black ; but they are concealed, and do not form the black band so conspicuous in *Clangula Barrovii*.   The specific appellation is intended as a tribute to Mr. Barrow's varied talents, and his unwearied exertions for the promotion of science.

" Head and two inches of the neck bright pansy-purple, with a greenish reflection on the ears ; forehead and chin brownish black.   Dorsal plumage, wings, and broad tips of the long flank feathers mostly velvet black.   Crescentic patch from the rictus to the sides of the forehead, lower part of the neck, shoulders, tips of the outer scapulars, lower row of lesser coverts, six secondaries, and under plumage pure white ; space round the thighs, the tail, and its lateral under coverts broccoli-brown ; bill blackish ; legs orange ; webs black.

" Bill shorter and narrower towards the point than that of the Golden Eye, and the feathers of the forehead, instead of running to a point on the ridge of the bill as in the latter, terminate with a semicircular outline. The plumage also of the occiput and nape is longer, forming a more decided crest than in that species. Wings two inches and a half shorter than the tail."

The female we have never seen, but have thought it best to append Dr. Richardson's description of that sex.

" Female—head and adjoining part of the neck umber-brown, without a white mark ; dorsal plumage pitch black ; its anterior part, particularly the shoulders and the base of the neck all round, edged with ash grey. A white collar round the middle of the neck.   Flanks clove-brown, edged with white.   Intermediate coverts blotched with white and black ; greater coverts white tipped with black secondaries as in the male.   Both mandibles orange at the point, their tips and posterior parts black.   Feet like the male."

Our Plate represents a male of the natural size.

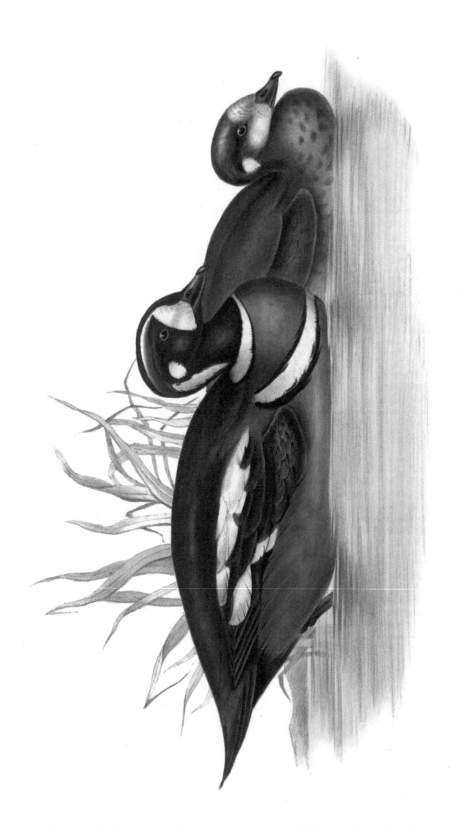

HARLEQUIN DUCK.

Anas histronica. *(Linn.)*
Clangula ——— *(Leach.)*

HARLEQUIN DUCK.

Anas histrionica, *Linn.*

Clangula histrionica, *Leach.*

NATURE, ever boundless in her resources and ever varying in her details, appears not unfrequently to delight in producing the most striking and singular contrast of colours, with which to adorn the plumage of the feathered race: and yet, strong and decided as the contrast may be, the whole effect is harmonious and delightful. It is so in the splendid bird before us, which gains its name from the multiplicity of its markings, —markings which, unlike those produced as it were by accident, and observed only upon varieties, without permanency and without method, are here the characters of a species, and are continued feather for feather through successive generations.

The general contour of its body together with the fanciful markings of its plumage would seem to ally it to the Mergansers; but on comparing the skeletons it was found by Mr. Yarrell and ourselves not to approach so nearly to that tribe of birds as did the skeleton of the Long-tailed Duck, *Harelda glacialis*, Leach. This circumstance, together with its general similarity in bone to the skeleton of the Golden Eye, has induced us to arrange the Harlequin Duck under the same genus with it.

Although higher latitudes than the British Islands constitute the true habitat of the Harlequin Duck, it has nevertheless been frequently captured here, and both sexes enrich several of our collections of native birds. Its presence, however, is attended with great uncertainty both in this country and in all the temperate portions of Europe. It is consequently prized by ornithological collectors as one of their greatest rarities. The Arctic regions, which are enriched with so many aquatic birds, afford this beautiful species a residence especially adapted to its solitary mode of life, as well as supplying it with every necessary conducive to its existence. It is said to be peculiarly local in its places of resort, preferring inland streams in the neighbourhood of waterfalls and cataracts. It is abundant in Norway, Russia, and Siberia; and was also observed by Dr. Richardson while pursuing his course from Hudson's Bay to the shores of the polar seas, sometimes in pairs, but not unfrequently in small flocks of eight or ten. It breeds near the water's edge, generally laying from six to twelve eggs, of a pure white. Its powers of diving are said to be great; hence it seeks its food, which consists of shell-fish and mollusca, at the bottoms of rivers and inlets of the sea. It flies with great rapidity, and is capable of performing extensive migrations in a very short time.

The disparity in size between the sexes is strikingly apparent, the female being full a third smaller than the male, and remarkably plain in her colouring. We have to thank the Rev. Dr. Thackeray, Provost of King's College Cambridge, for the loan of the female from which our figure was taken. That gentleman has also most liberally offered us any other species in his valuable collection for the same purpose.

The young male of the year very much resembles the adult female in colour; its superior size however, and enlarged windpipe, at once point out its sexual difference. They are at least four years attaining the fine state of plumage which characterizes the male in our Plate.

The cheeks and neck are black, with dark violet reflections; a triangular space between the beak and the eye, a spot behind the eye, a longitudinal band on the sides of the neck, two crescent-shaped collars on the breast, with parts of the scapulars, are all of a pure white; over each eye is situated a rufous band which terminates at the occiput; breast and belly blueish ash; flanks reddish chestnut; the whole of the upper surface deep blackish brown.

In the female, the upper surface is of a dark brown, with the exception of a white spot behind the eye, and a space of dull white which covers the face; the head lighter brown clouded with transverse markings of a darker colour.

We have figured an adult male and a female, about two thirds of their natural size.

LONG-TAILED DUCK.
Anas glacialis. *(Linn.)*
Harelda ——— *(Leach.)*

Genus HARELDA.

Gen. Char. *Beak* unarmed at its base, very short, slender, narrow towards the tip; the under mandible furnished with a small nail at its tip. *Nostrils* linear, basal. *Crown* elevated. *Tail* with its intermediate feathers elongated and pointed.

LONG-TAILED DUCK.

Anas glacialis, *Linn.*

Harelda glacialis, *Leach.*

Le Canard de miclon.

This species of Duck is generally diffused over the northern regions of the continents of Europe and America, but is to be considered more especially as a native of the Arctic circle, whence it diverges, but in diminished numbers, to more southern and temperate latitudes : it is, however, seldom found to extend its migrations in Europe further south than the British Islands. Wilson informs us, that in America straggling parties are found in winter as far south as Kingston in Carolina.

These birds resort to the Orkneys during winter, and a few remain there to breed during the summer. Of this fact we are well convinced, as we have ourselves received them in the breeding season in the height of their summer plumage. In severe winters the whole of our coast is visited by them in small parties, which seldom remain any length of time, but return, on the opening of the weather, to more congenial latitudes. M. Temminck says they are sometimes found in the great lakes of Germany and on the shores of Holland, but never in flocks. In Sweden, Lapland and Russia they are in great abundance. The nature of their food, which consists of mussels and other shelled and naked Mollusca, confines them almost exclusively to the sea, and they are remarkable for their activity and dexterity in diving. Wilson observes, that they are lively, restless birds, flying swiftly and sweeping round in short excursions.

This species of Duck builds a nest not unlike that of the Eider, and equally valuable for its down : it is placed among the grass and herbage which grow near the sea-shore, and is lined with down from the female breast. The eggs are from ten to fourteen in number, of a blueish white tinged with olive colour.

One of the most remarkable peculiarities in this bird is the contrast which exists between the plumage of winter and of summer, which may be thus described. The male in winter has the head and neck, with the exception of a blackish brown patch below the ears, of a pure white, as are also the scapularies and long pendent secondaries which fall over the quills ; the abdomen and outer tail-feathers are white ; the chest, back, shoulders, wings, and middle elongated narrow tail-feathers, which exceed the rest by several inches, blackish brown. In summer the white on the head which predominates in winter is exchanged for brownish black, which is then the most universal colour, except that the scapulars and tertials are reddish brown. Beak greenish black, with a transverse band of red ; tarsi and toes blueish lead colour ; membranes blackish ; irides orange.

The young males of the year and the adult females resemble each other. In both, the elongated scapularies and tail-feathers are wanting ; the top of the head and sides of the neck are brown ; the general colour of the whole of the upper surface dull brown, each feather having a darker centre ; the rest of the plumage white.

We extract the following description of the windpipe of this bird from Shaw's General Zoology, vol. xii. p. 177.—" The trachea of this bird is of a singular construction, and differs from that of the other *Anatidæ* : it rather increases in size at each extremity : at the lower end, close to the labyrinth, one side is flattened, and instead of the bony rings continuing round of their full breadth, this part is crossed with four distant linear bones as fine as a thread, which support a delicate transparent membrane three quarters of an inch in length, and almost three eighths of an inch broad at the base : below this ribbed membrane projects the bony part of the labyrinth, with a tympanum of a kidney shape placed transversely to the trachea, the middle of which is flat and membranaceous : the opposite side of the labyrinth is depressed, and from the bottom of this part the branchiæ take their origin."

The Plate exhibits a male and female, in their winter plumage, rather more than three fourths of the natural size.

WHITE-HEADED DUCK.
Undina leucocephala. (*Mihi*).

Drawn from Nature & on Stone by J & E Gould.

Printed by C. Hullmandel.

Genus UNDINA, *Mihi.*

GEN. CHAR. *Beak* elevated and protuberant at its base, with a furrow on the upper ridge; anterior half very much depressed and broad. *Nostrils* situated at the junction of the elevated and depressed portion. *Wings* very short, pointed, and concave; the first quill-feather the longest, the scapularies reaching to the end of the wing. *Tail* cuneiform, and composed of twenty narrow stiff feathers, unprotected at the base by coverts both above and below. *Legs* placed far behind. *Tarsi* flattened, and much shorter than the toes. *Toes* long, and fully webbed, the outer one the longest; hind toe situated high on the tarsus and lobated. General plumage dense, glossy, and adpressed.

WHITE-HEADED DUCK.

Undina leucocephala, *Mihi.*

Anas leucocephala, *Linn.*

Le Canard couronné.

ON attentively inspecting the genera comprising the European Ducks, as instituted by modern naturalists, it is evident that the present species is not referrible to any one of them at present recognised.

The contour of its form, the character of its plumage, and especially that of its graduated tail, composed of stiff elastic feathers, together with the large feet and the elevated position of the posterior lobated toe, indicate it to be exclusively aquatic in its habits; so near, in fact, does it approach the genus *Hydrobates* of M. Temminck, containing the Lobated Duck of New Holland, that we have no hesitation in asserting it to form the type of a closely allied genus.

The White-headed Duck is almost entirely restricted to the eastern countries of Europe, being very abundant in Russia, Poland, Hungary, and Austria; hence in Germany, France, and Holland it may be considered unknown. In its native countries it chiefly inhabits large sheets of saline waters and arms of the sea; and we are informed by M. Temminck, that so exclusively aquatic are the habits of this bird that its nest is even so constructed as to float upon the water, being composed of reeds and other water-plants. Although we have no account of its powers of diving, or mode of progression in its native element, we cannot for a moment doubt that it possesses every facility for making active and vigorous exertions in its congenial element. Its stiff elastic tail and waterproof plumage tend to support an opinion that, like the Cormorant and Darters, it swims almost entirely submersed beneath the surface, while its short concave wings and broadly webbed feet also assure us that in the power of diving it is second to none.

Its food is said to consist of molluscous animals and fishes.

The female differs from the male more in the obscurity of her markings and in the general dullness of her colouring than in any decided contrast, a circumstance in which we may again trace a resemblance to the New Holland Duck already alluded to.

Beak fine bluish lead colour; crown of the head black; forehead, cheeks, throat, and occiput pure white; chest, flanks, scapularies, and the whole of the upper part, of a fine reddish brown, transversely intersected with irregular zigzag lines of blackish brown; quills and tail black; under surface dull russet brown; tarsi and feet brownish; irides fine yellow.

The Plate represents an adult male of the natural size.

GOOSANDER.

Mergus Merganser *(Linn.)*

Genus MERGUS, *Linn.*

GEN. CHAR. *Beak* lengthened, nearly cylindrical, largest at the base, curved at the tip, nailed ; both *mandibles* armed at the edges with sharp angular teeth directed backwards. *Nostrils* one third from the base, oval, longitudinal, lateral, near the centre of the mandible. *Feet* webbed ; *outer toe* longest ; *hind toe* lobed and free ; *tarsi* compressed. *Wings* moderate ; second *quill-feathers* the longest.

GOOSANDER.

Mergus Merganser, *Linn.*

Le Grand Harle.

PRE-EMINENTLY distinguished by the breadth and boldness of its colouring, and the delicacy of some of its tints, the Goosander, both from its beauty and its superiority of size, is entitled to rank as the finest of its genus ; admirably adapted for diving, it possesses great power and agility in the water ; its flight also, when once fairly on the wing, is strong and rapid.

Its native locality appears to be the northern regions of the continents of Europe and America, where, among large and unfrequented lakes, it finds an asylum and breeding place : from these, its summer haunts, it migrates southwards on the approach of the severities of winter, seldom appearing in our latitudes unless the season indicates an extremely low temperature in the Arctic circle ; at such times it frequents our shores and unfrozen lakes, either in pairs or in small flocks of seven or eight : but the extensive inland waters of Holland and Germany appear to be its favourite place of resort.

The form of the body is long and compressed ; the total length twenty-six inches ; the weight four pounds.

Beak red on the sides, darker above, edges serrated ; armed at the end with an abruptly hooked nail. Head ornamented with slender elongated hair-like feathers, forming a voluminous crest of a rich glossy black with green reflections, which colour is continued half-way down the neck, where it terminates abruptly. The back and scapulars of a fine black ; wing-coverts and secondaries white. Quills blackish-brown ; rump and tail grey, the sides irregularly marked with fine waved freckled darker lines ; tail-feathers eighteen. The whole of the under surface of the body of a delicate yellowish cream-colour. Legs placed very far back. Tarsus and toes of a rich orange-red ; interdigital membrane rather darker.

The trachea presents two enlargements of the tube before it enters the labyrinth or inferior larynx, which consists of two irregular cavities divided from each other by a membranous partition.

The female is considerably less than the male, and differs from him not only in plumage but also in the anatomical structure of the trachea, which wants the enlargement both of the tube and the bony labyrinth. The beak, irides and feet are less brilliant in colour. Head, neck and crest rufous brown ; chin white; the uppersurface of the body uniform dark ash grey; the under part lighter with a tinge of cream-colour.

These differences in the female, connected with the similarity of plumage characterizing the young males of the year, (which are only to be distinguished by the masculine structure of the trachea, and rather larger size,) have induced early writers to consider them as a distinct species, to which they have applied the name of Dundiver ; an error corrected by more recent observation.

Its food consists of fish, small crustacea, and molluscous animals. Its flesh is rank and unpalatable.

The female is said to lay twelve whitish eggs, but the nidification of this bird is little known.

Our Plate represents a male and female in full plumage, two thirds of the natural size.

RED BREASTED MERGANSER.
Mergus serrator. (*Meyer*)

Drawn from life & on Stone by J & E Gould.

Printed by C Hullmandel.

RED-BREASTED MERGANSER.

Mergus serrator.

Le Harle huppé.

THE Red-breasted Merganser appears to be the only species of this genus which occasionally breeds with us, remaining the whole year in the Orkneys and about some of the inland lakes of North Britain, building its nest, which consists of dried bents, grass, &c., on any elevated situation, as a rocky bank near the water's edge ; and laying from eight to twelve cream-coloured eggs.

The British Islands appear to be the most southern limits of its summer abode ; but it is found in plenty on both continents within the arctic circle,—regions more congenial to its habits and more abundantly supplying its wants.

Its powers of swimming and diving equal if not exceed those of the other species of this genus ; its food is in all respects the same, and its flesh is equally rank and disagreeable.

The Red-breasted Merganser is one third less than the Goosander, which it resembles in its habits and manners, but differs from it extremely in colour. The beak is very long and slender, the sides red, separated by an upper line of black ; the head furnished with a crest of long, slender, recurved feathers, the whole of which, with a third of the neck, is of a dark glossy green ; below this a broad white band encircles the neck, gradually losing itself in the colour of the breast, which is of a chestnut-red, longitudinally blotched with dashes of black. The back and tertials are of a deep glossy black. On each side of the chest, overhanging the shoulders, is situated a singular tuft of broad and peculiarly formed feathers, the centre of each of which is occupied by a large white triangular spot, surrounded with a border of black ; the whole presenting a beautiful chequered appearance. The centre of the wing is white partly crossed with two slender bars of black. The quills are blackish brown. The sides and rump light grey elegantly marked with zigzag lines of black. Tail dark grey. The under surface of the body of a dirty white. The irides, legs, and feet, of an orange-red ; the webs darker.

During the period of incubation, however, the male undergoes a considerable change in plumage, losing the rich glossy green of his head and neck, which degenerates into an obscure brown, and the fine chestnut colour of his breast entirely disappears.

The female is rather less than the male, and exhibits in the rufous brown of the head, crest, and neck, one of the peculiarities of the genus. The beak and legs are duller than in the male ; the back and sides are grey ; the chest barred with obscure transverse spots ; the middle of the wings white, with a dark bar. The under surface of a dirty white.

The young male of the year resembles the female in colour, but possesses the characteristic conformation of trachea peculiar to the males.

We have figured an adult male and female, two thirds of the natural size.

HOODED MERGANSER.

Mergus Cucullatus *(Linne.)*

Drawn from Life and on Stone by J.&E. Gould.

Printed by C. Hullmandel.

HOODED MERGANSER.

Mergus cucullatus, *Linn.*

L'Harle couronné.

THE native locality of the Hooded Merganser appears to be the United States and the higher latitudes of North America. It is also found on the north and north-western coasts of Europe. We are indebted to that distinguished Ornithologist, Mr. Selby, of Northumberland, for a knowledge of the occurrence of this rare and beautiful species in England, and its consequent claims to a place in the Fauna of Great Britain. It is however but an accidental visitant, as this solitary instance only is on record of its having been taken in this country.—The following is the account given by that gentleman in the "Transactions of the Natural History Society of Northumberland, Durham and Newcastle," vol. 1. p. 292.

"The other (alluding to the present bird), which we may claim as an acquisition, is the *Mergus cucullatus* (Hooded Merganser), upon the authority of a specimen killed at Yarmouth, in Norfolk, in the winter of 1829. The skin of this individual was lately sent to me by my esteemed correspondent Mr. Elton, of Redland near Bristol, to whom it was presented by a friend, who purchased it as a rare variety in a *fresh state* from the person who actually shot it. From the state of its plumage it appears to be a young female, the crest not being so full or large, and the white upon the secondary quills less extended than in the skin of an adult female compared with it." We have not, it is true, examined the specimen from which Mr. Selby's figure and description were taken; nevertheless we are inclined to believe, from an inspection of the beautiful drawing which illustrates his work, that the bird in question is not a young female as Mr. Selby supposes, but an immature male, which in certain stages closely resembles the female in plumage, but may be distinguished by the larger and more rounded crest, which in the latter is long and thin.

In size the Hooded Merganser is intermediate between the Red-breasted Merganser and Smew, and partakes strongly of all the characters which are peculiar to the genus. The irides are golden; the bill elongated, narrow, and of a dull red; the head ornamented with a double row of long silky feathers, forming a beautiful compressed hood, which commences from the base of the beak, and when elevated forms a bold arch ending at the occiput. The head and its hood are of a glossy greenish black, with the exception on the latter of a large triangular fan-shaped spot of white the apex of which is situated just behind the eye, from which it diverges outwards, having its external edge bordered by a margin of black continued from the surrounding colour; the neck and back black; the chest white, with two beautiful crescent-shaped lines proceeding from the back and arching forward in a point near the centre of the chest; wings dark, with four alternate bars of black and white; quill-feathers brown; tertials consisting of elongated slender and pointed white feathers, with a broad black border, hanging gracefully over the wings; rump and tail dark umber; sides ferruginous brown, marked with minute undulating transverse lines of a darker colour; under surface white; feet and webs flesh-coloured.

The description of the female of a single species applies more or less to the same sex throughout the whole of the genus; as will be readily perceived in the present instance. Bill and feet as in the male; the head furnished with a small crest of slight hair-like feathers, of a dull ferruginous brown; the neck dusky brown, slightly barred towards the chest with white; the whole of the upper surface of a deep and rich umber; the wings bearing traces of the white bars, which are distinct in the male; the under parts white.

We have figured a male and female, two thirds of their natural size.

SMEW.

Mergus albellus, (Linn)

SMEW.

Mergus albellus, *Linn.*

La Piette.

THIS bird interests us more by the purity and contrast of its colours, than by its brilliancy or variety; the snowy whiteness of its plumage, broken by irregular markings of a jet black, in conjunction with the neatness of its general figure, producing an extremely pleasing effect.

The Smew is the smallest of the genus *Mergus*, and offers considerable deviation in some points from the typical form: the body is less compressed and elongated, the beak shorter, with feet of diminished size; which peculiarities, conjoined with the dark colour of the latter, evince a departure from the usual characteristics of this genus, and would seem to indicate an approximation to the more true *Anatidæ* or Duck tribe. Still however it must be conceded that it retains many prominent features of the genus, its habits and manners being in all respects the same; visiting us during inclement winters, at which time it is found in small numbers on the coasts and inland lakes of these Islands, and in much greater abundance in the more extensive waters of Holland and Germany; but, as far as observation goes, it has never been known to breed with us. The arctic regions of both continents seem to be selected for its summer residence and breeding place. The bill is shorter than the head, and tapers suddenly as it approaches the point; its general colour, as well as that of the feet and legs, is a blueish lead, the webs more dusky; irides dark hazel. The head is ornamented with a snowy pendent crest; a large greenish-black circle surrounds the eye; the occiput, over which hang the drooping feathers of the crest, is also black, with green reflections; the neck pure white, as is also the chest, on the sides of which two crescent-shaped lines of black bend forward, continued from the back, which is black, becoming grey towards the rump and tail. The scapulars are white, marked with oblique lines of black; the lesser coverts white, forming a broad band across the wing; secondaries and greater wing-coverts black, with white edges producing two smaller white bands; quills blackish-brown; tertiaries light blueish-grey; under parts pure white.

The trachea of this species differs from those of its congeners in having no enlargement of the tube, which is formed by a series of firm rings gradually increasing in size from the larynx to the labyrinth, the latter being small and irregular in its figure; the tongue also is rather more rounded, and in form approaches to that of the Ducks.

The female is smaller, and, although a pleasing bird, does not present the contrast of colouring so attractive in the male. The beak and legs are lead-coloured; the crest, as usual throughout the females of the genus, ferruginous-brown; a black oval spot below the brown extends from the base of the bill, covers the cheek, and surrounds the eye,—a marking which seems to have been entirely overlooked by European ornithologists. The occiput and neck are dusky-white; the chest light-grey, faintly clouded with obscure bars of a darker colour; the back, rump and tail deep greyish-brown; the wings, as in the male, are distinguished by a broad band and two lines of white; shoulders, quills, and tertiaries brownish-black; the under surface of the body white.

The young birds, both male and female, during their first winter are alike in plumage, but both want the black circle round the eye which distinguishes the adult female. Its weight is about twenty-four ounces; its length is from fifteen to sixteen inches: its food consists of marine insects, molluscæ, small fish, and water plants.

The annexed Plate represents an adult male and female in their winter dress, two thirds of their natural size.

GREAT CRESTED GREBE.
Podiceps cristatus. (Lath.)

Genus PODICEPS.

GEN. CHAR. *Bill* lengthened, strong, slightly compressed, straight, and pointed. *Nostrils* lateral, basal, linear, pierced in the middle of the nasal fosse; space between the corner of the bill and the eye naked. *Wings* short and concave. *Tail* none. *Legs* placed at the posterior extremity of the body, the *tibiæ* being inclosed within the integuments of the abdomen. *Tarsi* much compressed. *Feet* consisting of four toes, three before and one behind; the front ones much flattened, and each furnished with a broad membrane. *Nails* large, flat, and broad. *Plumage* soft, downy, and thick, with a silky lustre.

GREAT CRESTED GREBE.

Podiceps cristatus, *Lath.*

Le Grebe huppé.

THE *Podiceps cristatus* is not only the largest of all the European species of its genus, but may be regarded as one of the most typical examples. It is a native of the British Islands as well as of all the temperate portions of continental Europe, everywhere frequenting lakes, large ponds, the mouths of rivers, or the borders of the sea. In these situations it remains during the greater part of the year, eluding pursuit by its extraordinary powers of diving, and capability of remaining submerged beneath the surface of the water. We have also received numerous examples from Asia and Africa, which proved to be strictly identical with European specimens.

The Plate represents a young bird of the year, and an adult during the season of incubation, at which period it assumes the rich ornamental crest and tippet which are then so conspicuous. It is the immature bird which is described by the older writers as the Tippet Grebe, and which so nearly represents the adults in their winter dress, as to render any further description unnecessary; a mistake which modern ornithologists have rectified.

The full or red stage of plumage, in which the frill and crest appear, is not acquired until the third year, and even then, in its greatest luxuriance, is only the ornament of the season of pairing and incubation, the elongated plumes of the cheeks and head being lost, as we suspect, on the approach of winter. There is another peculiarity common to this and the rest of the Grebes which requires notice; we allude to the circumstance of the stomach being found after death commonly filled with a mass of feathers from the breast, but whether swallowed for the purpose of assisting the powers of digestion or not, it is impossible to conjecture. The nest is composed of masses of decayed aquatic vegetables, secured amidst the herbage on the margin of the water, with the variations of which it rises or falls. The eggs are three or four in number, of a greenish white stained with brown. Their food consists of fish, crustacea, and aquatic insects.

The plumage of this species may be thus described: Crown of the head and occipital tuft or ear-feathers deep greyish black; the frill black at its extreme edge, and rich chestnut throughout the greater part of the rest of its length, gradually fading off into the white of the cheeks and throat; the whole of the upper plumage brownish black, with a white bar across the wings; the under surface silvery white, becoming rufous on the flanks; the tarsus and toes dark olive green on the upper side, on the under side pale yellow; bill dark horn colour; irides red.

In winter the plumage resembles that of the summer, except that the richly coloured frill and elongated ear-feathers are wholly wanting. The sexes at either season offer no external differences in the plumage.

The figures in the Plate are somewhat less than the natural size.

RED-NECKED GREBE.
Podiceps rubricollis. (Lath.)

RED-NECKED GREBE.

Podiceps rubricollis, *Lath.*

Le Grebe jou-gris.

AMONG the European species of Grebes, the Red-necked is intermediate in size between the Crested Grebe (*Podiceps cristatus*) and the Horned Grebe (*Podiceps cornutus*). From the former it may be distinguished by the more partial development of the frill, which, with the whole of the cheeks, are of a light grey, and by the deep chestnut of the front and sides of the neck; while from the latter it differs in having the red streak passing through the eye to the occiput in the form of horns, as well as by the grey colour of the frill and cheeks, which in the *Podiceps cornutus* are black.

In point of rarity, particularly in our own climate, it is on an equal footing with the latter, being only an accidental visiter, though sometimes probably breeding with us.

In food, habits and manners this rare bird closely agrees with its congeners, inhabiting large inland lakes, rivers, estuaries, and the borders of the sea. In France and Holland it appears to be almost as scarce as in this country. It seems to be more common in Germany and Sweden; but its true habitat is the eastern portion of Europe or the adjacent regions of Asia. It is, however, far from being uncommon in all our larger collections; and we have ourselves seen both the young and adult in the London markets.

In their mature plumage, the two sexes offer but little external difference, both, we believe, always losing the beautiful frill and red colouring of the neck in winter, and regaining them early in the succeeding spring.

The young bird, when it has attained the full size, as in all this family, is of a greyish brown above, and white beneath, but may be distinguished by one character from the young of other species, namely, by the yellow colour of the whole of the lower mandible except at its very tip.

The adult colouring is as follows.

The top of the head, the egrets, the occiput and back of the neck, black; the back and whole of the upper surface of a brownish black, with the exception of the secondaries, which are white, so as to form a band across the wings; front and sides of the neck deep chestnut; under surface white; bill black, except at the base, which is of a rich orange yellow; irides scarlet; tarsi and toes dull olive green.

Our Plate represents an adult in the summer plumage, and a bird of the year, of the natural size.

HORNED GREBE.
Podiceps cornutus, (Lath.)

Drawn from Nature & on Stone by J.A.F. Gould.

Printed by C. Hullmandel.

HORNED GREBE.

Podiceps cornutus, *Lath.*

Le Grêbe cornu, ou Esclavon.

THIS beautiful Grebe is one of the rarest of those that occasionally visit the British Islands, particularly when it is adorned with the richly coloured plumes which characterize it in the adult state, a stage in which we have illustrated both sexes. We would however observe, that this fine plumage is only confined to the birds during the breeding season; at other seasons the plumage is less diversified, the frill and horns being alike absent, and the chestnut colouring of the neck being exchanged for greyish white; in this stage as well as that of the young it has been termed the *Podiceps obscurus, Podiceps caspicus,* &c. It is in the latter state that it is most frequently seen on our shores, and in which it so closely resembles the young of *Podiceps cristatus* as to require minute attention in order to distinguish them. To this difference we have alluded in the description of *Podiceps auritus.*

Of all the Grebes, the present has the widest range of habitat, extending itself throughout the whole of the arctic circle, seldom venturing further south than the British Isles in Europe, and the middle of the United States in the American continent. Like all other species of its genus, it seems to prefer inland lakes adjacent to the sea, and the mouths of large rivers; but still it is often found along low flat shores of the sea.

In its nidification it agrees strictly with its congeners, constructing a nest of such water-plants as abound on the spot in the lake where it resides; the nest being always on the surface of the water, attached to the strong reeds which rise from the bottom and secure it in its position. The eggs are four in number, of a dull white, exhibiting stains from the weeds upon which they repose.

The plumage of summer may be thus characterized. The top of the head, back of the neck, and upper surface black, with a slight tinge of green; a stripe of light chestnut which takes its origin from the base of the beak, and passes through the eye, is spread over two tufts of silky feathers, which rise like horns on each side of the occiput; the feathers of the cheeks are lengthened and spread out into a beautiful frill of a rich greenish black; the fore part of the neck and edges of the flanks of a rich chestnut; the under surface silvery white; tarsi and feet dull olive, with the exception of the anterior and posterior edges of the former, which are yellow; beak black, tipped with yellow; irides crimson.

The female strictly resembles the male, except that her size is rather less, the plumes less brilliant, and the ornament of the head less developed.

The Plate represents a male and female of the natural size.

EARED GREBE.

Podiceps auritus, (Linn.)

EARED GREBE.

Podiceps auritus, *Lath*.

Le Grêbe oreillard.

Like the rest of the family, whose locomotive powers are ill adapted for land, the Eared Grebe inhabits the water as its native element; not only obtaining its food there, but also carrying on the whole process of incubation, constructing a floating nest, composed of water-plants rudely matted together, which falls and rises with the influx and reflux of the waves. In size, the *Podiceps auritus* is somewhat less than the Sclavonian Grebe, or *P. cornutus*, from which it may readily be distinguished by the absence of the chestnut-coloured neck and rufous stripe which passes from the base of the bill through the eye to the occiput.

The present species may be considered as one of the rarest of the genus in this country; but we are led to suppose, from the seasons in which it is taken, that it sometimes resorts to our inland waters for the purpose of breeding. The female lays about four eggs, of a dirty white colour. The young differ very considerably from the adult; the characteristics of which we have faithfully portrayed in the annexed Plate, where it will be seen that the prevailing colour of the immature bird is a uniform grey on the upper surface, with a silvery appearance spreading over the whole of the under parts.

Reasoning from analogy, we may suppose that the *Podiceps auritus* undergoes the same variations at different seasons of the year which we know to take place in the other species of the genus; gaining its darker colour and ornamented ear-feathers only as the breeding season advances. When this period is past, and during the winter, we believe the adult to bear a close resemblance in plumage to the young of the year which have not yet undergone any change. The male and female offer but little difference.

M. Temminck informs us that it is extremely rare both in the marshes and on the coasts of Holland, its native locality appearing to be more especially confined to the rivers and fresh waters of the North of Europe.

Its food consists of small fishes, crustaceous animals, the larvæ of water insects, &c.

The bill is black; the irides bright red; the ear-feathers long and silky, radiating from the eye to the occiput, and of a light glossy chestnut; the head ornamented with a short full crest, which, with the throat, neck, and upper surface, is of a uniform blackish brown. The quill-feathers dark brown, secondaries white; sides of the rump dark chestnut brown. The whole of the under surface a pure silvery white; legs greenish-black.

Weight thirteen ounces; length twelve inches and a half.

LITTLE GREBE.

Podiceps minor, (*Linn.*)

Drawn from life & on Stone by J.& E. Gould.

Printed by C.Hullmandel.

LITTLE GREBE, or DABCHICK.

Podiceps minor, *Linn.*

Le Grêbe custagneux.

THE changes in plumage which even a bird so common as the Dabchick undergoes, have been until lately so little understood as to have produced for a single species a double nomenclature. It is now, however, known that the black-chinned Grebe of older authors, and by them supposed to be a distinct species, is the *Podiceps minor* in its summer plumage. In this state, as well as in that which it assumes in winter, we have introduced it in our Plate.

To this little bird, as to its congeners, the water is the native and familiar element. Extensively spread over Europe, except as we approach the more northern regions, it may be seen busily traversing the surface of inland waters, or dipping and diving in pursuit of its food; still it is shy, and distrustful of man, disliking his presence, and avoiding his prying curiosity by retreating at his approach to its reedy covert; or, if this be im-practicable, diving among floating weeds and water-lilies, where, with its bill alone above the surface, for the purpose of breathing, it will remain patiently watching till the danger be past, when it will cautiously emerge and seek its wonted haunt.

At ease and alert as is the Dabchick on the waters, it exhibits on *terra firma* a complete contrast, waddling along in an awkward and constrained manner, and glad to escape to its congenial element again. Its powers of flight are also inconsiderable; unless, indeed, it rises to a certain elevation, when, notwithstanding the shortness of the wings and absence of tail, it can sustain a long and rapid excursion.

The young when just excluded are in the perfect possession of all those powers which especially tend to their preservation. While yet covered with down, and perfectly incapable of flight, they may be seen, in com-pany with the parent birds, swimming and diving, either in the exuberance of animal enjoyment, or in pursuit of food : hence the legs and beak, which in most birds are long in acquiring their full development, outstrip in the present instance the acquisition of the powers of wing, this latter endowment being less immediately and intimately connected with their preservation and the manner of obtaining subsistence.

In winter, while in its brown plumage, the Dabchick gives the preference to broad extensive waters, lakes or rivers, associating together in small numbers during the season of clouds and storms, and on the approach of spring separating by pairs in different directions over the country in search of a more secluded and con-genial breeding-place; often taking up their abode in small ponds close to the habitation of man.

Should the lover of nature watch them at the time of their building without being discovered, (an attain-ment of the utmost difficulty, as every sense seems alive to danger and gives notice of intrusion,) he will be delighted to mark their playfulness and agility, while the tone of happiness and enjoyment which pervades their actions and their mutual labours, throws an additional charm over the picture.

The nest of this bird is composed of a mass of green plants, loosely interwoven, which floats on the water. In this the female deposits her eggs to the number of four or five, the original whiteness of which becomes discoloured, apparently from the juices of the plants in contact with them, and the wet feet of the parent birds.

The female Dabchick, at least under certain circumstances, (as the author has often personally witnessed,) is undoubtedly in the habit of covering her eggs on leaving the nest; and he has watched while this action has been performed : it is effected in a rapid and hurried manner, by pulling over them portions of the surround-ing herbage.

In the summer plumage the beak is blackish ; tip, base, and naked skin which extends to the eye, yellowish-white ; eyes reddish ; crown of the head, back of the neck, and chin, of a brownish black with green reflections ; sides and front of the neck of a lively chestnut ; the whole of the upper surface together with the sides and wings blackish with olive reflections ; the thighs and rump tinged with ferruginous ; the under surface more or less silvery ; legs and feet dark olive-green, without assuming a flesh-colour on the inner surface.

In winter the colour of the plumage differs little from that of the birds of the first year, which consists of a uniform brown above and more or less silvery beneath ; the two sexes having little external distinction either in winter or summer.

The total length nine inches.

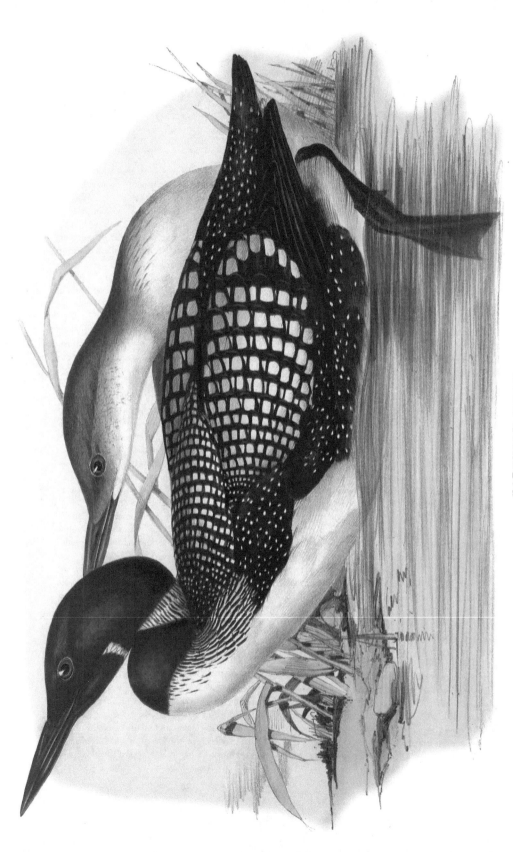

NORTHERN DIVER.
Colymbus glacialis (Linn.)

Genus COLYMBUS, *Lath.*

GEN. CHAR. *Bill* longer than the head, strong, straight, compressed, and sharp-pointed. *Nostrils* basal, lateral, linear, oblong, pervious, half closed by a membrane. *Wings* short, acuminate, having the first quill-feather the longest. *Tail* short and rounded. *Tarsi* thin, very much compressed. *Feet* large, of four toes, three before and one behind; outer toe exceeding the middle one in length; front toes entirely webbed; hind toe furnished with a lobe, and partly connected with the exterior membrane of the inner toe. *Nails* flat and broad.

NORTHERN DIVER.

Colymbus glacialis, *Linn.*

Le Plongeon Imbrim.

THIS noble species of Diver, the largest and most typical of its genus, appears to be equally dispersed over the whole of the northern hemisphere, giving preference to the regions within the arctic circle during summer, and progressing southward as far as latitude 36° on the approach of autumn and winter, at which seasons they are by no means of rare occurrence in our own islands, although, in accordance with that general law of Nature which causes the young to wander furthest from their native habitat, we find a much greater proportion of immature birds than of those which bear the beautifully contrasted livery of the adult. The great dissimilarity of plumage which characterizes the species at different ages has caused much confusion in its nomenclature, the bird of the first year having been described as specifically distinct from the adult, while, again, those of the second year, when they have partially assumed the mature livery, have been considered as differing from both. In this latter state they are frequently met with round our coasts, but less so than the birds of the first year, which may be very commonly observed even in our lakes and estuaries.

M. Temminck states that on the European Continent it gives a preference to the shores of the sea, although the young are frequently found up the large rivers; and even the German and Swiss lakes are not altogether exempt from its visits. That it inhabits the shores of the Black Sea, and without doubt the Mediterranean, is a fact with which we are made acquainted by the circumstance of the Zoological Society's having received an individual, in the second year's plumage, from their valued correspondent Keith E. Abbot, Esq., of Trebizond.

We have before alluded to the far-extended wanderings of the young migratory birds; and as no instance is on record of these birds having bred in southern latitudes, we may reasonably infer that the individual above mentioned had wandered from the regions of the arctic circle, which form their almost exclusive summer residence and breeding-place. Are we, then, to presume that the individuals found in the Black Sea have proceeded thither by way of the Atlantic Ocean and Mediterranean Sea, or by crossing partially over land, following the course of the large rivers, as the Danube, the Don, the Volga, &c.? We incline to the latter supposition, as all migratory animals pursue, with as little deviation as possible, a course from north to south, or *vice versâ*.

Its dependence for food rests entirely on its great activity in diving, as it subsists solely, whether at sea or in fresh water, upon fish, aquatic insects, &c., in the capture of which it displays astonishing agility and rapidity of motion.

The situation chosen for the purpose of nidification are the borders and islands of inland seas, lakes, and rivers; the nest being placed close to the water's edge, so as to be easily accessible to the parent, whose perfect adaptation for progression on the water, its natural element, totally unfits it for walking, though it contrives to propel itself forward by means of resting its breast upon the ground and striking backward with its feet, somewhat like the action of swimming.

The plumage of the sexes is strictly similar, the adults having the top of the head and neck fine black glossed with purplish green; a transverse bar of white spotted with black crosses the throat, and a wider band of the same colour passes lower down upon the back part of the neck; the whole of the upper surface glossy black, each feather having white spots, one on each side the shaft, forming rows, those on the scapularies becoming larger and square, but continuing small and nearly round on the back and rump; primaries black without spots; flanks and sides black spotted with white; whole of the breast and under surface white; bill and legs black; irides reddish brown.

The young of the year has the top of the head, back of the neck, upper surface, and flanks light greyish brown, the centre of each feather being darker; under surface pure white; bill, inner side of the tarsi, and interdigital membrane fleshy greyish white; outer side of the tarsi and toes brownish black.

The Plate represents an adult male and a young bird of the year, about two thirds of the natural size.

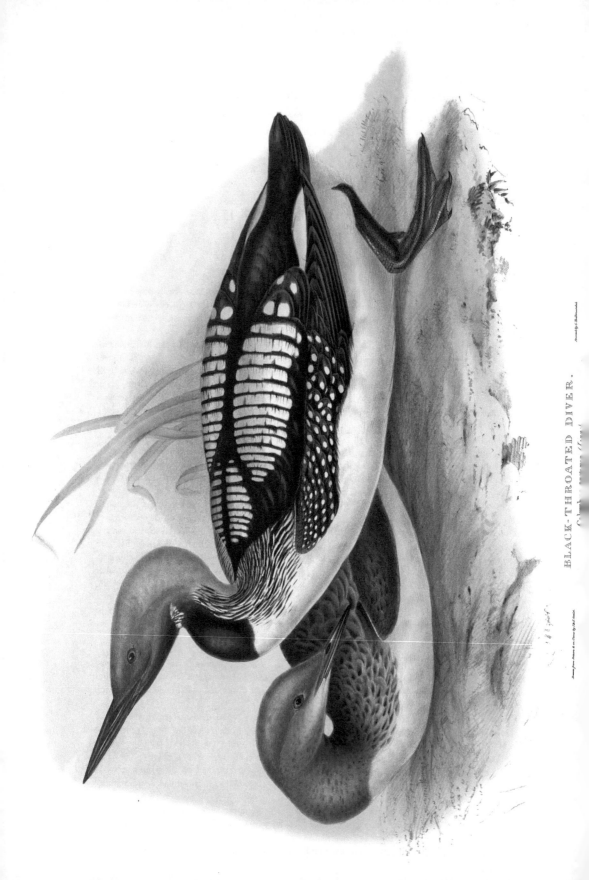

BLACK-THROATED DIVER.

BLACK-THROATED DIVER.

Colymbus arcticus, *Linn.*

Le Plongeon Lumme, ou à gorge noire.

THE accompanying figures of the young and adult of the Black-throated Diver were taken from two specimens of exquisite beauty, placed at our disposal by W. Baker, Esq., of Bayfordbury in Hertfordshire, by
whom they were shot during his tour through Norway in the pleasant pursuit of natural history. That
gentleman informed us that they were both killed upon one of the small lakes of the interior, from which circumstance we may infer that it was an adult, with its own young of that year.

The range of the Black-throated Diver extends over the whole arctic circle, everywhere giving preference
to inland waters and small lochs. It must have afforded much pleasure to Sir Wm. Jardine, Bart., and Mr.
Selby, during their late visit to the extensive wilds of Sutherlandshire, to observe a pair of these birds inhabiting almost every small loch they visited. Before this period it had not been fully ascertained that this species
made any portion of the British Islands a permanent residence, or that it remained in them during the period
of incubation. It is of rare occurrence in all the temperate portions of the globe, its migrations being less
extended, perhaps, than any other species of its genus, the young, as is generally the case, wandering furthest
from home. It is not an unfrequent occurrence to find individuals in their first year's plumage in the London
market. While in this stage they often frequent the sea and the mouths of large rivers, where they obtain an
abundant supply of fish, crustacea, and other marine animals, which constitute their principal food. It will
be scarcely necessary to inform our readers, that when the inland lakes of the northern climes become frozen,
the adults retreat to the ocean, where they brave with impunity the severities of the coldest winters. Their
power of diving is vigorous and remarkable, not more for swiftness than for the long time they are capable
of continuing submersed : it is this amazing power that enables them to avoid with the greatest ease every
artifice of man to capture them on the open sea or large lakes.

When fully adult the sexes offer little or no difference in the colouring of their plumage : the young, on the
contrary, are clothed in a much more sombre vest, being entirely destitute of the black throat and contrasted
bars of white and black which ornament the back and scapularies of the adults.

The nest is placed at the brink of the water; the eggs being generally two in number, of a dark olive
brown blotched with spots of black.

Head and neck dusky grey, deepest on the fore part of the head; throat and front of the neck deep black
with purple and green reflections ; below the throat a narrow transverse band of black ; a broad band, longitudinally rayed with white and black, extends from the ears down each side of the neck ; upper surface deep
glossy black ; the greater part of the scapulars and the side feathers of the mantle marked with large white
spots, forming on the scapulars several transverse bars ; wing-coverts spotted with white ; sides and flanks
black ; the remainder of the under surface being pure white ; legs deep brown on the outsides, paler within ;
bill brownish black ; irides reddish brown.

The young undergo three moultings before they attain their full colouring, during which changes they
gradually pass from the plumage of the year represented in our Plate to that of the adults.

The Plate represents a male in full plumage, and a young bird of the year, about three fourths of the
natural size.

RED-THROATED DIVER.
Colymbus septentrionalis; (Linn.)

RED-THROATED DIVER.

Colymbus septentrionalis, *Linn.*

Le Plongeon à gorge rouge.

THE Red-throated Diver, although possessing all the characteristics of the genus, differs very considerably both in colour and disposition of its markings from the other two species, viz. the Northern, and Black-throated, which form its European congeners. Of these it is the least in size, but by far the most abundant, being found in considerable numbers along the European coasts, and is especially plentiful in the arctic regions of both the European and American quarters of the globe. Although far from being uncommon on the sea-shores, it appears to evince great partiality for fresh and inland waters, taking to the sea when these are frozen ; hence in winter it is common along the coasts of England and Holland, beyond the latitudes of which it rarely passes southwards.

Like the others of its genus, it is a bold and vigorous diver, a mode of progression to which its lengthened neck and body are admirably adapted. Its principal food consists of small fish, crabs and other crustacea, to which, when it visits fresh waters, are added frogs, newts, and aquatic plants. Its flight, when it is mounted into the higher regions of the atmosphere, is rapid, and it has the capability of sustaining it for a long time.

We are able from experience to assure the reader, that the British Islands, particularly Scotland, the Orkneys and Hebrides, are to be reckoned among its annual breeding-places ; nor have we any reason to doubt that the northern shores of Europe in general afford it an asylum also for a similar purpose. It constructs a slight nest of grasses and vegetable fibres, among the herbage of morasses, and at the edges of such large lakes as invite it by their seclusion and quietude. Its eggs, two in number, are of a dark red brown blotched over with spots of black. The young, immediately after exclusion from the shell, are very active, follow their parents to the water, and instantly commence their search for food. Their growth is rapid, so that they soon acquire their full size, though not the adult colouring of the plumage till after the first autumn.

The young bird has hitherto been described in works on Ornithology as the *Colymbus stellatus*, the Speckled Diver, &c. ; the white throat which characterizes the birds of the year, together with the speckled plumage of the upper parts, having doubtless led to the mistake. The throat and fore part of the neck in the adult bird are covered with short, thick-set, velvety feathers of a dark grey, having in the centre a broad longitudinal band of rich chestnut,—at least during the summer, for we are as yet in doubt whether this character be lost as winter advances ; certain it is, that the adult birds taken in autumn have both the grey and chestnut of the throat less distinct and intermingled with patches of white. When the bird has become fully adult, the white specks which more or less distinguish the upper plumage of the young birds, according to their age, disappear, leaving the back of a uniform greenish brown. The upper part of the head and back of the neck are marked with longitudinal, wavy, narrow lines of white edged with black ; breast and under parts white ; flanks dashed with ash-coloured blotches ; beak black ; irides orange ; tarsi on their external aspect, and toes, of a deep olive black, having their internal sides and webs of a livid white.

We have figured an adult and a young bird of the year two thirds of their natural size.

FOOLISH GUILLEMOT.
Uria Troile: (Lath)

Drawn from Nature & on Stone by J & E Gould.   Printed by C Hullmandel.

Genus URIA.

GEN. CHAR. *Beak* of mean length, straight, strong, compressed, and pointed ; upper mandible slightly arched ; *tomia* intracted ; angle of the lower mandible gently ascending ; *commissure* nearly straight. *Nostrils* basal, lateral, concave, longitudinally cleft, and half covered with the feathers of the *antiœ*, or projecting side angles of the cranium. *Wings* short, narrow, and acute. *Tail* of twelve or fourteen feathers, very short. *Legs* situated at the back of the abdomen and concealed within its integuments. *Tarsi* short and compressed. *Feet* of three toes, all directed forwards and palmated ; outer and middle toes of equal length, the inner one much shorter. *Claws* fulcate, the middle one the longest.

FOOLISH GUILLEMOT.

Uria Troile, *Linn.*

Le Grand Guillemot.

THE native habitat of this well-known bird extends throughout the northern regions of both hemispheres, and it is probably more abundant than any other of the oceanic birds of the same family. In its habits and manners it is truly aquatic, making the sea its permanent place of residence, except during the season of breeding, when it assembles on the ledges of the precipitous rocks which overhang the deep. In the British dominions, it collects annually in vast multitudes at the high cliffs and the Needle rocks in the Isle of Wight, the Bass rock in the Firth of Forth, the steep rocks on the west and east coast, the Shetlands and Orkneys, and, in fact, any coast whose bold and precipitous rocks afford a place suited for incubation. Here, united with Puffins, Cormorants, Razorbills, and Gulls, the assembled multitude offers to the eye a striking, novel, and animated picture, their continued clamour mingling with the hoarse beating of the sea, and completing a scene of no ordinary interest to the lover of nature ; the Puffin seeking his hole in the rock, the Cormorant and Shag resorting to the topmost cliff, and the broad-winged Gull the lowermost range, covered with scanty herbage, while the Guillemot takes possession of the middle ledges along the face of the precipice, where thousands may be seen patiently performing the work of incubation, each sitting upright on its single large egg, which were it not for its peculiar shape would every moment be swept away from its narrow resting-place. After the breeding-season is over, they again take to the watery element, where, with their young, they traverse the wide ocean, not returning to the rocks till the succeeding spring. After breeding, they appear to undergo a partial moult : they lose their primaries so simultaneously as to be incapable of flight for a considerable period ; a circumstance of little moment, as they easily elude pursuit by diving, in which they excel surprisingly. At this time they also begin to lose the obscure black of the cheeks, which is exchanged for delicate white ; this white also characterizes the young of the year, which can only be distinguished from the adult in winter, by the darker colouring of the body, and by the more abbreviated and fleshy coloured bill. There is no perceptible difference in the sexes at either season.

Independently of the great resort of this bird to the British Isles, they abound in similar situations along the whole of the coast of Northern Europe, whence they gradually migrate southwards on the approach of winter, returning again with the vast shoals of fishes which pass northward in spring.

The whole of the upper surface of the throat and neck is of a uniform sooty black, inclining to grey, with a slender bar of white, which extends half across the wings ; the under surface is a delicate white ; bill blackish brown ; irides dark hazel ; tarsi dark brown, with a slight tinge of olive.

The Plate represents an adult and a young bird of the year of the natural size.

BRIDLED GUILLEMOT.
Uria lachrymans; *(Lapyl.)*

Drawn from Nature & on Stone by J. & E. Gould.

Printed by C. Hullmandel.

BRIDLED GUILLEMOT.

Uria lacrymans, *Lapyl.*

Le Guillemot bridé.

ALTHOUGH we have figured this bird under the name of *lacrymans* we are doubtful of its specific value, bearing as it does so close a resemblance to the common species (*Uria Troile*), and from which it differs only in the white mark which encircles the eyes and passes down the sides of the head. It inhabits the same localities, and is even often found in company with the common species, and that too on various parts of our coast, part ticularly those of Wales, where, we have been informed, both kinds are equally numerous. It was firs-described as distinct by Choris, who states that it is abundant at Spitzbergen and the neighbouring seas. By M. Temminck and the French naturalists the two birds are considered to be distinct, and as such we have figured them.

The head, neck, all the upper surface, wings, and tail are deep sooty black; a line encircling the eye and passing down the side of the head, the tips of the secondaries, and all the under surface pure white; bill black; feet brown.

Our figure is of the natural size.

BRUNNICH'S GUILLEMOT.
Uria Brunnichi, (Sabine)

Drawn from Nature & on Stone by J.& E. Gould.

Printed by C. Hullmandel.

BRUNNICH'S GUILLEMOT.

Uria Brunnichii, *Sabine.*

Le Guillemot à gros bec.

This species may at all times be distinguished from the *Uria Troile*, with which it has been often confounded, by the stout and abbreviated form of the bill, and by the much shorter space between the nasal orifices and the tip. We are only able to give a figure of it in its supposed summer plumage, which resembles that of the Common Guillemot, with this exception, that the dark parts are deeper and much more intense in colour, verging indeed towards sooty black. Although we have no authentic account of its having been captured in the British Islands, we feel convinced that it must occasionally occur along our northern shores. It is very abundant in Greenland, Spitzbergen, Davis's Straits, and Baffin's Bay; it doubtless also tenants the rugged shores of Norway, Lapland, &c.; and its habits in all probability closely resemble those of its congeners.

Crown of the head and all the upper surface glossy brownish black; sides of the face and front of the neck dark sooty black; tips of the secondaries and all the under surface white; bill black; gape bright yellow; feet greenish.

Our figure is of the natural size.

BLACK GUILLEMOT.
Uria grylle, (Lath.)

Drawn from Nature & on Stone by J.& E. Gould.

Printed by C.Hullmandel.

BLACK GUILLEMOT.

Uria Grylle, *Lath.*

Le Guillemot à miroir blanc.

The northern parts of Scotland and the Orkney and Shetland Islands form a place of general rendezvous for the Black Guillemot, which being less migratory in its habits than its near ally the Foolish Guillemot (*Uria Troile*, Lath.) seldom quits these isolated groups, whose bays afford it shelter during the stormy season of winter, and whose abrupt and precipitous cliffs are equally inviting as a site for incubation. On the ledges of these rocks it deposits its single white egg spotted with black: the young are hatched in about three weeks, and shortly after are conveyed, but by what means is unknown, to the water, an element to which they are so expressly adapted that they are enabled to swim and dive with the utmost facility the moment they arrive on its surface, and to brave with impunity the rough seas which are so prevalent in northern latitudes. After the process of reproduction is over, the adults are subject to a considerable change in the colour of their plumage, apparently caused by a general moult, even to the primaries, which are so simultaneously lost that the bird is for a considerable period deprived of the power of flight. The fine black plumage by which the Black Guillemot is characterized during the summer now gives place to a mottled dress, consisting of half white and half black feathers unequally dispersed over the body, the former colour predominating so much during the rigorous season of winter as to render the bird almost wholly white. Although this style of plumage characterizes, to a certain extent, the young of the year, still the latter may at all times be distinguished from the former by having the tips of each feather, which is white beneath, only fringed with black; by having the white spot on the wings, at all times uniform in the adult, invariably clouded with black; and by the feet being yellowish brown instead of red.

Although a few pairs of the Black Guillemot occasionally breed on the Isle of May in the mouth of the Frith of Forth, still it is evident that the high northern latitudes form its most congenial and natural habitat. It appears to abound in the arctic circle, being equally common in the polar regions of both continents. According to the continental writers, it is less abundant on the coasts of Holland and France than on those of England, but more frequent on those of Norway and the shores of the Baltic. It rarely, if ever, resorts to inland waters.

Its principal food consists of small fish, marine crustacea, &c.

The sexes are alike in plumage, and the adults, in summer, may be distinguished by their having the whole of the plumage of a sooty-black tinged with olive-green, with the exception of a snow-white patch on the centre of each wing; bill black; irides and feet red.

The Plate represents an adult and a young bird of the natural size.

GREAT AUK.
Alca impennis. (Linn.)

E. Lear del.

Drawn from Nature on Stone by J.B.E.smith.

Printed by C.Hullmandel.

Genus ALCA.

GEN. CHAR. *Beak* straight, arched, large, very much curved at the point; both mandibles laterally sulcated, and covered for half their length with short feathers. *Nostrils* near the lower edge of the upper mandible partly concealed by feathers. *Wings* short, narrow, and in one species unequal to the purpose of flight. *Legs* short, situated far backwards. *Toes* three before, webbed as far as the claws; hind toe wanting; front of the tarsi and toes scutellated. *Tail* short, pointed, and consisting of twelve or sixteen feathers.

GREAT AUK.

Alca impennis, *Linn.*

Le Pingouin brachiptère.

IN this noble species of Auk we recognise a close approximation to the true Penguins, which form the genus *Aptenodytes*: being, like them, destitute of the powers of flight, its narrow slender wing serves more as an oar for aquatic progression than for any other decided purpose; unless, perhaps, in assisting the bird to scramble up the rocks, on the ledges of which it deposits its single egg, which is, indeed, the only time at which it makes the solid earth its abode.

The seas of the polar regions, agitated with storms and covered with immense icebergs, form the congenial habitat of the Great Auk: here it may be said to pass the whole of its existence, braving the severest winters with the utmost impunity, so that it is only occasionally seen, and that at distant intervals, even so far south as the seas adjacent to the northernmost parts of the British Islands. It is found in abundance along the rugged coasts of Labrador; and from the circumstance of its having been seen at Spitzbergen, we may reasonably conclude that its range is extended throughout the whole of the arctic circle, where it may often be seen tranquilly reposing on masses of floating ice, to the neighbourhood of which in the open ocean it seems to give a decided preference. Like the common Razorbilled Auk, it exhibits an annual change in the colours of its throat and neck, the jet black of these parts giving way to white in winter. Deficient as the Great Auk may be in the powers of flight and of easy unconstrained progression on the land, these deficiencies are amply compensated by its extraordinary capability of diving and its express adaptation to the watery element: here it is truly at ease, following its prey and sporting in the midst of the waves. Its food consists exclusively of fish of various species, which, however rapid they may be in their motions, it captures with the utmost facility.

Its single egg is deposited on the naked rock, either in some natural fissure or crevice just above the reach of the highest tides; its colour white tinged with buff, marked with spots and crooked lines of brownish black. The young take to the water immediately after exclusion from the egg, and follow the adults with fearless confidence.

There exists but little or no difference between the size or plumage of the sexes. In summer the whole of the upper surface is black, with the exception of a large white space before the eyes and the tips of the secondary quill-feathers; the whole of the under surface white; bill and legs black, the former being marked with oblique transverse furrows of a lighter tint.

The Plate represents an adult in its summer dress about two thirds of the natural size.

RAZOR-BILL AUK.

Alca Torda, (Linn.)

Drawn from Nature & on Stone by J. & E. Gould.

Printed by C. Hullmandel.

RAZOR-BILLED AUK.

Alca torda, *Linn.*

Le Pingoin macroptère.

THE habits and manners of the Razor-bill so closely approximate to those of the Common Guillemot (*Uria Troile*, Linn.), that the same description equally applies to both; to enter into them fully would therefore be only repeating what we have said in our account of the last-mentioned bird: like it, the Razor-bill inhabits the wide expanse of the ocean, the severities of which it braves with the utmost indifference; indeed it appears to rejoice in the agitation of the billows, that brings around it multitudes of small fish, which constitute its only support; like it, the Razor-bill, when called upon by the impulse of nature to the great work of incubation, seeks the inaccessible cliffs round the coasts of our island, on which it assembles in immense flocks, to deposit each its single egg on the barren ledges of the rock; and so often do the eggs of the two species resemble each other, that they are scarcely to be distinguished except by a practical observer: that of the Razor-bill is somewhat less, and generally has neither the grotesque marking nor the deep green colour which characterize the greater portion of the eggs of the Guillemot. The Razor-bill is very generally distributed throughout the seas of the arctic circle, a portion of the globe of which it is more especially a native; never, we believe, extending its migrations beyond the temperate latitudes of Europe in the Old World, and the southern portions of the United States in the New. In point of numbers the Razor-bill does not appear to equal its ally, if we may judge by what is to be observed along our own shores: the Guillemots literally swarm during the breeding-season on most of the rocky shores not only of our island but of the northern portions of the Continent in general. The dissimilarity which exists in the beak of the young from that of the fully adult Razor-bill has been the source of no little confusion, and has given rise among ornithologists to synonyms which were erroneously bestowed as specific titles on the young of the year, before the bird had been duly developed, a circumstance which does not take place until the second year: this mistake was further strengthened by the total absence of the white line between the eye and the beak, in birds whose size is equal to that of adults. It is, however, a singular fact, that when just excluded from the egg, this white line is strikingly apparent on the down with which they are then clothed; but with the acquisition of the feathers, this white line disappears, and is regained with the stripes on the upper mandible towards the close of the second year.

During winter the adults of both sexes lose the dusky colouring of the throat precisely in the same manner as the Guillemot. At this period the old and young closely resemble each other in plumage, and are only to be distinguished by the character of the beak.

The sexes are alike in colouring.

The whole of the upper surface and the throat is of a deep sooty black; a distinct white band crosses the wing, and a white line passes from the upper part of the bill to the eye; the remainder of the plumage is white; the bill is black, the upper mandible marked with deep transverse furrows and a clear white band; feet and tarsi brownish black.

Our Plate represents an adult and a young of the year, of the natural size.

LITTLE AUK.

Alca alle; *(Linn.)*
Mergulus alle; *(Bon.)*

Drawn from Life & on Stone by J & E Gould.

Printed by C. Hullmandel.

Genus MERGULUS.

Gen. Char. *Beak* medial, its base furnished with downy feathers, somewhat thickened, above convex, emarginate towards the tip, curved. *Nostrils* rounded, half-covered with feathers. *Legs* short, three-toed, webbed. *Wings* short.

LITTLE AUK.

Alca alle, *Linn.*

Mergulus alle, *Bon.*

This interesting little oceanic bird, which we have illustrated in the accompanying Plate, inhabits the intermediate countries extending northwards from our latitude to the borders of perpetual ice, occurring equally in the polar regions of both continents. In these severe and high latitudes, it congregates in almost innumerable flocks. Their numbers are often diminished by the crews of vessels, as well as by the native Esquimaux; their flesh being considered both wholesome and delicate, at the same time affording a beneficial change of diet. They are said to be very tame and easily captured,—a circumstance readily accounted for, as the persons engaged in the whale fisheries, and the limited race of natives inhabiting the borders of these seas, are the only human beings they are ever disturbed by.

In these wild and almost impenetrable regions, the Little Auk, it will be observed, finds an almost secure asylum and breeding-place, as well as an element congenial to its habits and mode of life; and it is only from extreme necessity, chiefly from the severities of winter, that it seeks, for a short period, an asylum in more temperate climes. Its visits to the British Isles, and Europe in general, therefore, must be considered more as an accidental occurrence than a periodical migration. Young birds are, as is the case with the young of most species, found to wander furthest from their native habitat; the examples, therefore, we obtain, as well as those from Holland, France, and Germany, average about ten young birds to one adult. We have been particular in our inquiries as to whether the Little Auk breeds in any of the northern Isles, being induced to believe so from the circumstance of a specimen now and then coming to hand in the season of incubation, and in its mature state of plumage: we have not, however, been able to collect any certain data by which to set the question positively at rest; and it yet remains for some zealous naturalist to supply the information. Like the rest of its family, the Little Auk passes a great portion of time on the ocean, where it sports with great ease and fearless self-possession, feeding upon marine insects, small crustacea and fishes, diving for its prey with great celerity and adroitness. Although the sexes offer little or no external difference, still the plumage undergoes considerable changes periodically, which we have illustrated in our Plate. The bird represented in the drawing with a black throat, is in its summer plumage; at this season, the whole of the head, neck and upper surface being black, with the exception of a white band across the secondaries; the scapulars are bordered with the same, and a small spot of white also appears over each eye; the breast and under surface pure white. In the winter plumage, as well as in the young of the year, the throat, like the rest of the under surface, is pure white; beak black; legs and feet of a brownish yellow.

The egg of this species is 1 inch 7 lines long by 1 inch 1 line wide, of a uniform pale blue, very similar in colour to the eggs of the Starling.

We have figured the birds of their natural size, in summer and winter plumage.

PUFFIN.
Mormon Fratercula. *(Temm)*

Drawn from Life & on stone by J. & E. Gould.   Printed by C. Hullmandel.

Genus MORMON, *Temm.*

Gen. Char. *Beak* shorter than the head, of greater depth than length, and very compressed; both *mandibles* arched, furrowed, and notched at the point, the edge of the upper one acute and elevated at its origin. *Nostrils* lateral, marginal, linear, naked, almost wholly closed by a naked membrane. *Tarsi* short, retiring. *Feet* palmated. *Toes* three before only, the two inner nails much hooked. *Wings* short; first and second *quill-feathers* equal, or nearly so.

PUFFIN.

Mormon Fratercula, *Temm.*

Le Macareux moine.

Whenever Nature appears to indulge in excentricity in the modification of those organs which are essential to existence, we are not to suppose,—because we cannot follow her through all her mysteries, or discover the motive,—that she ever acts a blind or random part, and the more inclined shall we be to come to this decision, the more closely we examine her operations. We make these remarks, because in the singular construction of the bill of this bird we are at a loss to account for this deviation from the forms which we see possessed by birds whose food and manners are altogether the same.

On a first glance at the Puffin, we cannot fail to be struck by the short and inelegant contour of its figure, and by the strange shape but brilliant colour of its beak, which imparts a singular aspect to the physiognomy of this inhabitant of the ocean; where, as if to belie its round and awkward figure, it displays great agility and an arrow-like quickness of motion;—its beak, deep, compressed, and pointed with a sharp ridge and keel, affords the beau ideal of an instrument for cutting through the water,—a circumstance the more necessary when considered in connexion with a form of body by no means so well adapted for diving with ease and vigour as is possessed by many others. The feathers however are thick, close and smooth, so completely throwing off every particle of water as to render it impossible that the plumage can be wetted. Independent of the use of the beak as a water-cutting instrument, it is a weapon of destruction to innumerable hordes of fry and smaller fishes which swim near the surface of the water. These, at least during the breeding season, are retained by dexterous management between the mandibles, till a row of little pendent victims is arranged along each side, their heads firmly wedged in the beak, and their tails and bodies hanging outside. Thus loaded, as we have frequently ourselves witnessed, the Puffin flies home to its mate or newly hatched offspring. The young, however, are themselves very soon ready for the water, where, long before they are capable of flight, they may be seen in company with their parents diving and sporting on their congenial element. The old birds evince great regard for their young, attending them with assiduity and manifesting the utmost anxiety on the approach of danger.

The Puffin has an extensive range, abounding at the season of incubation on the rocky parts of our coasts, especially the high cliffs and pointed rocks of the Isle of Wight, the rugged and precipitous coast of Wales, Scotland, the Orkneys and Hebrides, as well as on the northern shores of the European and American Continents. It does not however invariably resort to crevices and ledges of rocks, but is known occasionally, during the breeding season, to inhabit deserted rabbit-burrows, or other holes in the ground adjacent to the sea, within which, without constructing any nest, it deposits one or two eggs of a uniform dull white. The young when hatched are covered with long and fine down of a sooty black: their bill, as might be expected, is not so fully developed either as to size or colour, nevertheless it bears the characteristic peculiarity so striking in the adult bird; the sides also of the mandibles do not possess the deep furrows which appear at more advanced age. As they grow up, the general plumage assimilates to that of the adult, but is more obscure in its markings. The adult birds present no external sexual differences, the colour of the whole of the upper surface being of a dark brown, inclining to black with coppery reflections; a collar of the same colour passing round the neck; the cheeks white, shaded towards the lower parts with delicate gray; the whole of the under surface white; the bill blueish ash at its base passing off to a bright reddish orange, with three oblique furrows on the upper and two on the lower mandible; the gape covered by a naked puckered membrane; irides blueish gray; rim round the eye orange; above and below the eye, on the edges of the eyelids, are small bodies of a horny consistence and a deep slate-colour, that below the eye being narrow and two lines in length; their use is not known; legs orange. Total length eleven to twelve inches. Their food consists of fishes and marine insects.

Our Plate represents a male and female in different positions, to exhibit the peculiar character of their physiognomy.

NORTHERN PUFFIN.
Mormon glacialis, *(Leach).*

Drawn from Nature & on Stone by J & E Gould.   Printed by C. Hullmandel.

NORTHERN PUFFIN.

Mormon glacialis, *Leach*.

Le Macareux glacial.

To a superficial observer the *Mormon glacialis* would appear to present but little difference from the common species, so plentiful during the breeding-season on many of the rocky coasts of our island; but on a comparison of the two species we feel convinced that our readers will coincide in our opinion, of their being specifically distinct. Its larger size and more powerful bill, which is of a uniform rich orange colour, together with the greater length of the fleshy appendages over the eyes, will at once serve to distinguish this species from its near ally the *Mormon fratercula*.

The Northern Puffin, as its name implies, is almost strictly confined to the ice-bound regions of the arctic circle, over the whole of which we have reason to believe it is distributed, numerous specimens having been from time to time brought home from Spitzbergen and Kamtschatka by our navigators on their return from exploring the boreal regions.

It occurs but rarely in the temperate latitudes, hence both the coasts of America and Europe are only occasionally visited by it; the northern shores of Russia, Lapland, &c. being among the only places in our quarter of the globe where it may be looked for with certainty.

The sexes do not appear to differ in the colouring of their plumage, which may be thus described:

Crown of the head and occiput light brownish grey tinged with lilac; collar surrounding the neck, all the upper surface, wings, and tail brownish black tinged with blue; primaries blackish brown very slightly margined with paler brown; sides of the face and all the under surface white; bill bright orange red; corrugated skin at the angle of the mouth gamboge yellow; irides orange red; irides and horny appendages grey; legs and feet orange with the webs paler and the claws yellowish brown.

We have figured an adult of the natural size.

Genus PELECANUS.

GEN. CHAR. *Beak* longer than the head, long, straight, broad, much depressed; the upper
mandible channelled, the tip bent down and unguiculated; the lower bifurcate, furnished
with a flaccid dilatable sac. *Face* and *throat* naked. *Nostrils* basal, placed in a longitudinal cleft. *Legs* strong, short; the tibiæ naked at the base; the middle claw with its
inner edge entire.

PELICAN.

Pelecanus Onocrotalus, *Linn.*

Le Pélican blanc.

THOSE of our readers who are desirous of seeing this noble bird in a state of nature need only pay a visit to
the southern and eastern portions of Europe to gratify their laudable curiosity. Although the tropical climates of Africa and India constitute its natural habitat, nevertheless the eastern rivers of Europe, such as the
Danube and Volga, the extensive lakes of Hungary and Russia, and the shores of the Mediterranean, are places
in which it dwells in abundance. It is a species strictly confined to the Old World, over a great portion of
which it is plentifully distributed. M. Temminck, who quotes the *Pelecanus Philippensis* as synonymous
with this bird, states that individuals sent him from Egypt and South Africa do not differ from those taken
in Europe.

The *Pelecanus Onocrotalus* is a very large species, measuring nearly five feet in length, and from tip to tip
of the wings, when spread, about twelve or thirteen feet, and is remarkable both for longevity and for the long
period requisite for the completion of its plumage. The first year's dress is wholly brown; the feathers of the
back and breast being broad and rounded. The assumption of the lanceolate feathers, and the rosy tints that
pervade the plumage of the adults are only acquired as the bird advances in age; and, judging from individuals
which we have opportunities of noticing in a state of partial confinement, it would appear that a lapse of five
or six years is required before it may be considered fully mature.

The food of the Pelican is fish of all kinds, in the capture of which it displays considerable activity and cunning; and, although its robust body and immense bill would seem to contradict the assertion, its motions are
so quick that even young fry and eels can scarcely escape its vigilance. The same power that renders it so
light and buoyant on the water denies it the means of diving, consequently it is only the small and shallow
inlets of rivers that are resorted to when in search of its food. Occasionally, however, it will rise to a considerable height in the air, from whence, on perceiving a fish, it descends with astonishing swiftness and unerring aim, the rapidity of its descent forcing it beneath the surface, on which its extreme lightness causes it
to reappear instantaneously.

The female constructs a nest on the ground, of coarse reedy grass, generally about a foot and a half in
diameter, and lined with soft grass, laying two or more eggs, which are white, and like those of the Swan,
During the period of incubation it is extremely assiduous in procuring food for its young, carrying it to them
in the capacious pouch or gullet which forms so conspicuous a feature in this bird, and which is "one of the
most remarkable appendages that is found in the structure of any animal. Though it contracts nearly into
the hollow of the jaws, and the sides to which it is attached are not (in a quiescent state) above an inch asun-
der, it may be extended to an amazing capacity; and when the bird has fished with success, its size is almost
incredible: it will contain a man's head with the greatest ease. In fishing, the Pelican fills this bag, and does
not immediately swallow his prey; but, when this is full, he returns to the shore to devour at leisure the fruits
of his industry." (Shaw's General Zoology, vol. xiii. Part I. p. 111.)

The Pelican bears confinement remarkably well, particularly if space sufficient be allowed for it to plume
and wash itself. Fish either dead or alive are voraciously devoured; and fish in plenty obtained, it appears
content and satisfied.

Although it possesses the power of perching on trees, yet it gives the preference to rocky shores, which
appear to form the best and most natural situation when in a state of repose. On level ground its walk is
awkward and inelegant, and when on wing its flight is heavy and apparently effected with great labour.

Along the top of the upper mandible runs a line of crimson, the remainder being reddish at the base and
yellowish at the tip; the under mandible pale red; the pouch reddish yellow; the naked spaces around
the eyes are flesh colour; the occiput is slightly crested; the whole of the plumage is white, tinged more or
less with salmon colour, with the exception of the crest and a few pendulous feathers attached to the lower part
of the neck, which are pale yellow, and the primaries and spurious wings, which are black; legs flesh colour;
claws grey; irides hazel. The salmon-coloured tint which pervades the whole of the plumage is considerably
heightened during the breeding-season.

We have figured an adult male about one third of the natural size.

DALMATIAN PELICAN.
Pelecanus crispus (Bruch.)

DALMATIAN PELICAN.

Pelecanus crispus, *Feld.*

A BIRD of such striking magnitude as the present having so long escaped observation even on the shores of Europe, what may we not expect from those more distant countries to which the scrutinizing eye of the naturalist has seldom penetrated? Although this species has been introduced to the notice of the scientific within the last few years only, it has doubtless long abounded where it is now found. The specimen from which our figure is taken was sent us by Baron de Feldegg, and was one of twenty-four killed by him on the shores of Dalmatia.

In the letter which accompanied this specimen the Baron thus writes : " The first example of this bird that came under my notice was shot by myself in the year 1828 in Dalmatia, and was sent to the Imperial Cabinet in Vienna. Two years after this, Messrs. Rüppell and Kittlitz met with this species in Abyssinia, where, however, it would appear to be very scarce, as those gentlemen procured only a single specimen. In the year 1832 I published a description of it under the name of *Pelecanus crispus*. Many ornithologists are of opinion that there is only one species of Pelican in Europe, for which reason they have given it the specific name of *onocrotalus*, and they observe that the size which the bird attains is regulated by the temperature of the climate in which it resides. I possess examples of the true *Pelecanus onocrotalus* taken in Europe and at the Cape of Good Hope, which in all the more important points closely resemble each other; the tarsi, for instance, are of equal length, and the naked spaces round the eyes are of the same extent, while, on the contrary, one shot in Moldavia was much smaller. The *Pelecanus crispus* has undoubtedly escaped notice in consequence of no other naturalist having seen both species together as I have in Dalmatia, where it arrives in spring and autumn, and where it gives preference to the neighbourhood of Fort Opus on the river Naranta, which is bordered with morasses. It comes through Bosnia, seldom alone, but generally in flocks; I have seen as many as twelve together hunting for fish : it is very cunning, and is extremely difficult to shoot. I obtained, at different times, as many as twenty-four examples."

The *Pelecanus crispus* differs from the Common Pelican in possessing a beautiful crest and mane of narrow, elongated, silky feathers; in the naked space around the eye being smaller; in the feathers of the breast being stiff, lanceolate, rounded at the points, and of a firm elastic texture; in the body being more bulky and larger in all its proportions; and in the tarsi being stouter, of a different colour, and considerably shorter. "At all seasons of the year old birds may be found both with and without the crest. I saw a specimen in M. Aker's menagerie which had always had it, while a bird of the same species in another menagerie had none; in all other respects they were the same, and were very healthy; and I possess a female in which the ovaries were largely developed, and which has a large crest covering the whole of the head, which circumstance induces me to conclude that it is a very old bird."

Of its habits, manners, mode of nidification, &c. no details have been ascertained, but we may reasonably suppose that in all these particulars it closely resembles the other members of its family.

" Naked space round the eyes reddish, towards the bill bluish; upper mandible grey, passing into blue and red; gular pouch or sack under the bill blood red intermingled with bluish; feet bluish grey; head furnished with a crest and thickly covered with feathers, which, with the whole of the upper and under surface, are silvery white;" the tail is composed of twenty-two feathers, the shafts of which, with those of the scapularies and secondary wing-coverts, are black; primaries blackish brown; chest tinged with pale yellow.

The young, which are very seldom seen, are wholly brownish grey, the feathers being much finer and closer in texture and more silky in appearance than in the adults.

We cannot close our account of this splendid bird, one of the noblest of its race, without offering our warmest thanks to our esteemed friend the Baron de Feldegg for the very fine specimen of this bird he so obligingly sent us, as also for the very interesting communication transmitted with it, from which are extracted the paragraphs in this paper included between inverted commas.

The Plate represents an adult and a young bird about one fourth of the natural size.

COMMON CORMORANT.

Genus PHALACROCORAX, *Briss.*

GEN. CHAR. *Beak* generally longer than the head, straight, strong, hard, slightly compressed, having the upper mandible terminating in a powerful hook, and furrowed laterally from the base as far as the tip of the lower mandible, with the terminating nail distinct; its tip compressed and truncated; *tomia* of the lower mandible retracted. *Nostrils* basal, concealed. *Face* naked. *Throat* dilatable. *Wings* moderate, the second quill-feather the longest. *Tail* moderate, rounded, and composed of stiff elastic feathers. *Legs* placed far behind, short and strong. *Feet* of four toes, all connected by a membrane; outer toe the longest, the others gradually shortening to the hind one. *Tarsi* reticulated; the upper part of the toes scutellated.

COMMON CORMORANT.

Phalacrocorax Carbo, *Steph.*

Le Grand Cormoran.

OUR Plate illustrates this fine although common species in its nuptial dress, a style of plumage which it does not possess in a perfect state above one month out of the twelve, and the peculiarity of which state consists in the narrow white feathers which ornament the sides of the head and neck, together with an occipital crest of long slender black feathers down the back of the neck, and a patch of pure white on the outer side of each thigh. This conspicuous plumage, which is common to both sexes, is assumed about the latter end of February or the beginning of March, the period at which these birds commence the work of nidification, after which the white plumes, together with the white patch on the thighs, gradually disappear, these parts then becoming of a uniform blueish black. This peculiarity of plumage is only found in birds that have attained their third or fourth year, the immature dress up to that period being of a dull brown colour, while the young of the year have the under surface wholly white. These remarkable and contrasted changes have induced the older naturalists to look upon individuals in the various stages of their existence as constituting so many distinct species; but further observation has fully proved their identity, the difference depending upon age and season.

The Common Cormorant is equally and rather numerously distributed along the coasts of the British Islands, often resorting to inland lakes and rivers adjoining the sea, especially such as are not liable to be frozen during winter. They occasionally perch and roost on trees, towers, and rocky projections; and although the summits and ledges of rocks overhanging the sea are the principal and favourite breeding-stations, still it is known to incubate occasionally in trees, and even upon the ground, as is the case in the Farn Islands, and the extensive reed-beds in Holland. The nest is usually composed of dried sea-weed, rudely put together, and often of a considerable thickness; the eggs, generally three in number, are of a greenish white, covered with a chalky coating, and extremely small compared with the size of the bird.

In swimming, the body of the Cormorant is nearly all emersed below the surface of the water, the tail serving as a very effectual rudder, by means of which it is able either to dive or turn in the most rapid and dexterous manner.

Its food, as may naturally be supposed from its powers and structure, consists almost wholly of fish, which it takes by chasing beneath the surface, the dilatability of its throat enabling it to secure and swallow fish of comparatively large dimensions; and we may easily conceive that the quantity it devours and the destruction it occasions in the shoals at various seasons of the year must be enormous, and injurious to the interests of the fishermen. Its distribution over Europe is in the same ratio as in the British Isles, and it is even more abundant on the rocky coasts of the north.

The plumage of spring :—On the back of the head are long plumes, which form a crest of slender feathers of a deep glossy green; on the throat extends a collar of pure white; on the top of the head, and on a great part of the neck and on the thighs, are long silky plumes of pure white; the feathers of the back and wings are of an ashy brown, bronzed in the middle with a broad edging of glossy greenish black; quills and tail-feathers black, which is the general colour of the under surface; beak dull white, clouded and transversely rayed with black; naked skin of the face greenish yellow; irides bright green; tarsi black.

In winter the top of the head, the neck, and thighs entirely lose the white plumes of spring, and are of the greenish black of the rest of the under surface.

The young have the top of the head and upper surface deep brown, with greenish reflections; the whole of the under surface white, more or less clouded with brown according to age.

Our Plate represents an adult male in the spring plumage, and a young bird of the year, three fourths of the natural size.

BLACK CORMORANT.
Phalacrocorax graculus. (Meyer).

Drawn from Nature on Stone by J & E Gould.

Printed by C. Hullmandel.

BLACK CORMORANT.

Phalacrocorax Graculus, *Briss.*

Le Cormoran nigaud.

OUR knowledge of the identity of this species is due to the kindness of M. Temminck, who favoured us with a fine specimen, from which the accompanying figure was taken. It is one of those species respecting which much doubt has hitherto existed, and which has led to considerable confusion in the works of our later British ornithologists, who have confounded it with the Common Shag or Green Cormorant. Notwithstanding all our endeavours, we have never been able to substantiate its claim to a place among the British Fauna; a circumstance somewhat singular, as M. Temminck gives Holland, together with a wide range through the northern latitudes of both worlds, as within the limits of its habitat. His words are these: "Habitat, The northern and meridional portions of both worlds. It is migratory in the eastern countries of Europe; is less numerous in its passage in the countries bathed by the ocean; and is very abundant in the regions of the artic and antarctic circles." In addition to this, he adds, that he has received individuals from Africa and North America which differ in nothing from those killed in Holland; and still further, that those killed in Brazil, of which he has seen a great number of examples, differ in no respect from those taken on his own coast. As for ourselves, with the exception of one specimen from Newfoundland, which we believe to be identical with the present bird, we have never seen it among the many extensive collections which it has fallen to us to examine, from the localities assigned by M. Temminck as its habitat. From the Shag and the Cormorant, the only two well-authenticated examples of this genus natives of the British Islands, there are abundant grounds of distinction, as an examination of the colouring and other generic characters will prove.

The figure in our Plate is that of a bird in mature plumage, in the act of assuming the characteristic marks of the breeding-season, which consist of a multitude of delicate white linear dots, occupying the sides of the head and neck and the feathers of the thighs, which moreover are accompanied by a thick occipital crest of black feathers, assumed at the same time with the delicate markings. In size it is inferior to the Cormorant, but rather larger than the Shag, from both of which it may be distinguished, among other things, by the lancet-shaped feathers of the back and scapularies, and by the yellowish red colour of the gular pouch. In habits and manners it is identical with its congeners, of which we may take the Common Cormorant as the most familiar example. Its food is fish, which it takes with the utmost dexterity by diving.

The head, neck, middle of the back, and all the underparts black; wings and scapularies brownish ash in the middle, with a margin of jet black; beak reddish horn colour, black along its upper margin; naked skin round the eye and gular pouch reddish yellow; irides brown; tarsi black. As the breeding-season approaches, the occiput becomes adorned with a crest of glossy black feathers, and the sides of the head and neck and the thigh-coverts become sprinkled with the linear dots of white to which we have already adverted.

The Plate represents an adult bird, about three fourths of its natural size, assuming the white dots and crest of the breeding-season.

LITTLE CORMORANT.

Carbo pygmæus. (Temm.)

LITTLE CORMORANT.

Carbo pygmæus, *Temm.*

Le Cormoran pygmée.

THIS species, although termed *pygmæus*, is very far from being the least of its genus; still it is much less than any other of its European relatives. It is the eastern portion of Europe alone which constitutes its true habitat, being very common in some parts of Hungary, and especially along the borders of the lower Danube: it is more rare in Austria, and is seldom seen in Germany. From the tracts it frequents in Europe, we are naturally led to expect that it is distributed over the adjacent portion of Asia, and we learn that it is found in great numbers in Asiatic Russia.

In the periodical changes of its plumage, and also in the changes which occur in its progress from youth to maturity, it strictly resembles the Common Cormorant; the birds of the first year having the usual brown tint pervading the upper surface, and the mottled greyish white on the chest and lower parts: as they advance to maturity this dress give place to a more decided and glossy plumage of jet black and grey. At the pairing-season, like the Common Cormorant, it becomes temporarily decorated with numerous fine linear feathers, of a white colour, on the sides of the head, neck, and thighs. Although we have every reason to believe that both sexes participate in this change, we cannot positively assert that this is the case.

The adult male in summer has the whole of the plumage of a glossy greenish black; each feather of the back and wings margined with black; neck, head, and thighs ornamented with fine filamentous white feathers; the rest of the plumage black.

The winter dress resembles that of summer, except that the fine white feathers on the head, neck, and thighs are entirely wanting.

The Plate represents a male of the natural size, undergoing the change from winter to summer.

SHAG OR GREEN CORMORANT.

Phalacrocorax cristatus. (Nivch Flem.)

Drawn from Nature & on Stone by J. & E. Gould.

Printed by C. Hullmandel.

SHAG or GREEN CORMORANT.

Phalacrocorax cristatus, *Steph. and Flem.*

Le Cormoran largup.

THE present species is rather abundantly distributed throughout the rocky and precipitous shores of the British Islands ; indeed it is yet a question whether it does not exceed in number the larger species, *Phalacrocorax Carbo*, with which it associates, particularly during the breeding season, but from which it may easily be distinguished, as also from all the other Cormorants, by the intense green of its body, and, during the season of incubation, by the elegant flowing semi-erect crest, and by the absence at this period of the delicate white markings on the sides of the neck and thighs, so conspicuous in the other species of the genus. The crest is only assumed during the season of reproduction, and is not found at all in birds of the first and second year. When fully adult, the sexes offer no difference in their external characters. In habits and manners they strictly resemble the Cormorant : like that bird, they may be observed going from their craggy haunts early in the morning, either out to sea or up the large rivers in quest of their prey, and regularly returning in small companies on the approach of sunset ; thus strongly reminding one, in these particulars, of the habits of the Rook, which almost every person must have observed going out in the morning, and returning in the evening to its accustomed roosting-place.

The Shag is widely distributed over all the northern portions of Europe, and as far southward as the shores of the Mediterranean. Its powers of flight are very great ; but not more remarkable than its powers of diving and continuance beneath the surface, where it makes its progress by repeated strokes of the pinions aided by its broadly webbed feet : in this way it easily secures its prey, often fish of large dimensions, for the carrying of which its dilatable throat is well adapted.

The site chosen for incubation is the topmost ledge of some bold precipice, where, secure from the interruption of man, it constructs a nest of dried sea-weed for the reception of its eggs, which are two, three, or four in number, of an oblong shape and a white colour, with a rough calcareous surface. When first excluded from the egg, the young are quite naked, but quickly become covered with thick black down, which remains a considerable time before it is succeeded by the regular feathers.

The adults have the whole of the head, neck, centre of the back, and under surface fine dark green ; the scapularies and wings bronze green, each feather being bordered by a narrow band of velvet black ; quills and tail black ; bill blackish horn colour ; angles of the mouth, skin round the eyes and gular pouch fine yellow ; feet black ; irides green.

The young of the year is distinguished by the whole of the upper part of the plumage being brown, slightly tinted with green ; and the under surface brownish ash, more or less inclining to white.

The Plate represents an adult in the summer plumage, and the young of the year, about three fourths of their natural size.

DESMAREST'S CORMORANT.
Phalacrocorax Desmarestii.

DESMAREST'S CORMORANT.

Phalacrocorax Desmarestii.

Le Cormoran de Desmarest.

The present bird is a native of the rocky shores of the eastern parts of Europe, or more properly speaking of the shores of the Black Sea and its tributary streams, and in these localities would appear to represent our common Shag, a species to which it very closely approximates both in size and general appearance, but on a comparison of the two birds from these different localities, no doubt can exist as to their being specifically distinct. The present bird, although not inferior in the size of its body, is decidedly superior in the length of the wing, while it possesses a much longer and more attenuated bill. Our specimens were received from M. Temminck, but no account of its habits and manners has been transmitted to us; we have, however, every reason to believe them to be the same as those of our native species.

The head, which is slightly crested, and the neck are greenish black; the whole of the upper surface green, each feather having a narrow margin of jet black; rump glossy greenish black; abdomen sooty black tinged with green; naked space at the back of the bill rich orange; bill yellowish horn colour; primaries, tail, and feet black.

We have figured an adult of the natural size.

SOLAN GANNET.
Sula Bassana, (Briss.)

Ed. Lear. d.

Genus SULA.

GEN. CHAR. *Bill* longer than the head, thick, strong, straight, acuminate, compressed towards the point, with the dertrum slightly convex; mandibles equal; the upper one laterally sulcated from the base to the tip, and with a hinge near the posterior part, making it appear as if composed of five separate pieces; culmen rounded; lower mandible having the angle rather prominent, gently ascending to the tip; chin-angle narrow and long, filled with a naked dilatable skin; face naked; tomia intracted, obliquely and unequally serrated. *Nostrils* basal, concealed from view. *Wings* long and acuminate. *Tail* graduated. *Legs* abdominal. *Tarsi* short. *Feet* of four toes, all connected by a membrane; the middle and outer toes of nearly equal length; middle claw having its inner edge dilated and toothed.

SOLAN GANNET.

Sula Bassana, *Briss.*

La Fou blanc ou de Bassan.

THE seas bordering the European shores are the natural habitat of the Solan Gannet, and nowhere is it more common during summer than on the rugged and precipitous coasts of Scotland, especially the Bass Rock, the isles of Ailsa, St. Kilda, &c., where they breed in vast multitudes: these situations, in fact, appear to be the principal nursery for this race. On the approach of autumn they leave their rocky breeding-places, and go further out to sea, the greater number passing considerably to the south, feeding on herrings, pilchards, and other fishes. Being destitute of the power of diving, they seize their prey by a vertical plunge when within a certain distance, and so forcible is their descent, that we are informed of instances in which they have killed themselves, by darting at fish attached to a board, connected by a rope fifty or sixty yards in length to a vessel at anchor, the neck being either dislocated or the bill driven firmly into the wood. The flight of the Gannet is extremely rapid, vigorous, and capable of being long sustained; hence it traverses the wide surface of the ocean with comparative ease. Although in the breeding-season the Gannets congregate in such countless multitudes, it seldom happens that they are found in flocks out at sea, but mostly alone, dispersed apparently in search of food. On the approach of spring they return to their accustomed rendezvous, which is generally preserved from molestation and farmed by persons who make a profit of the feathers and the young birds, their flesh being considered by some a delicacy, though to most persons its oily and fishy flavour renders it extremely disagreeable. While sitting on their nest, which is composed of sea-weed and other similar materials, they are so absorbed in the task of incubation, that they will sometimes permit themselves to be approached, and even handled, without quitting it: they lay but a single white egg, which in size is between those of a Cormorant and a Common Goose. From the time it is hatched till it arrives at maturity, no bird undergoes a more marked change in the colouring of its plumage. Mr. Selby was informed by two persons who rent the Bass Rock that the Gannet is four years in attaining a permanent state of plumage, and until this period has elapsed it is not known to breed.

The first year's plumage is characterized by the head, neck, and all the upper surface being blackish grey, inclining to brown, each feather tipped with a triangular spot of white; the breast and under surface white, each feather being edged with greyish black; the quills and tail greyish black, the shafts of the latter being white; the bill blackish grey tinged with brown; the irides pale brown; the legs and feet deep grey.

The second year's, by the head and greater part of the neck being white, more or less spotted with blackish grey; the upper surface of a nearly uniform brown, the white spots on the tips of the feathers becoming less distinct or entirely disappearing; and by the under surface becoming whiter.

The third year's, by the white increasing all over the body; the scapulars and tertials remaining black or spotted with blackish grey.

The fourth year's, or fully adult plumage, is characterized by the crown of the head, occiput, and upper part of the neck being pale yellow; all the remaining plumage pure white, with the exception of the quills and bastard wing, which are black; bill bluish grey, fading into white at the tip; naked skin round the eyes greyish blue; membrane at the gape and beneath the throat black; irides pale yellow; streak down the front of the tarsi and toes bluish green; webs blackish grey; claws greyish white.

The Plate represents a fully adult bird, and a young one of the first year, about three fourths of the natural size.

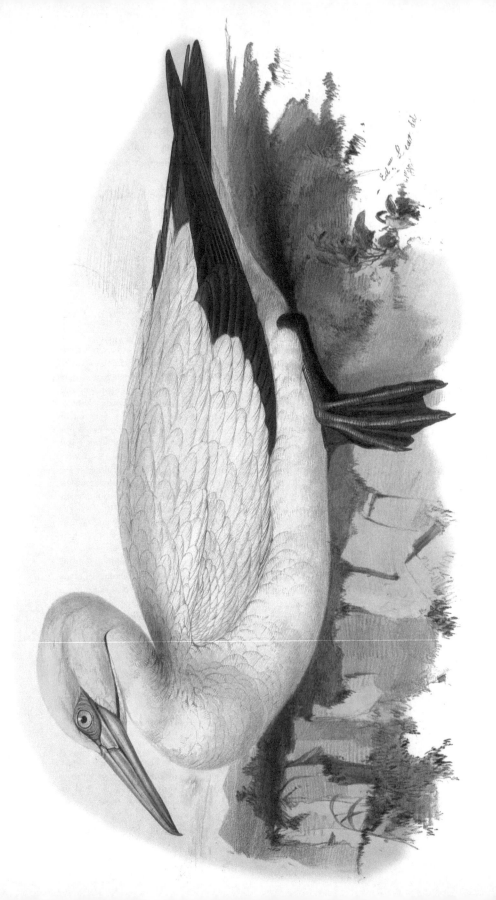

E. Lear del et lith.

BLACK-TAILED GANNET.
Sula melanura. (Temm.)

Printed by C. Hullmandel.

BLACK-TAILED GANNET.

Sula melanura, *Temm.*

WE are indebted to the kindness of M. Temminck for the loan of the fine specimen of this bird from which the accompanying figure was taken, and which he informed us was killed in Iceland. In every respect, except in having a black tail, it resembles the Solan Gannet, *Sula Bassana*, which bird when fully adult has a white tail. Whether this difference be an accidental variation, or if not, whether the difference is of sufficient importance to justify a specific distinction, we are unable satisfactorily to decide. M. Temminck regards it, we believe, as a true species, and as such we publish it, leaving the question still open for further investigation: at all events it will not be destitute of interest to the scientific naturalist. It may perhaps be said that as the Gannets change from almost black to white in passing from youth to maturity, this bird exhibits an intermediate state of plumage, the original black colouring still remaining on the tail: there can be no doubt, however, that the bird was fully adult, and we may remark that the first change that occurs in the plumage of the young shows a disposition in all parts to approach towards the colouring of the adults; it is therefore unlikely that the tail should be even of a deeper black than that of the young Gannet, while the rest of the plumage is that of complete maturity.

The figure is about three fourths of the natural size.

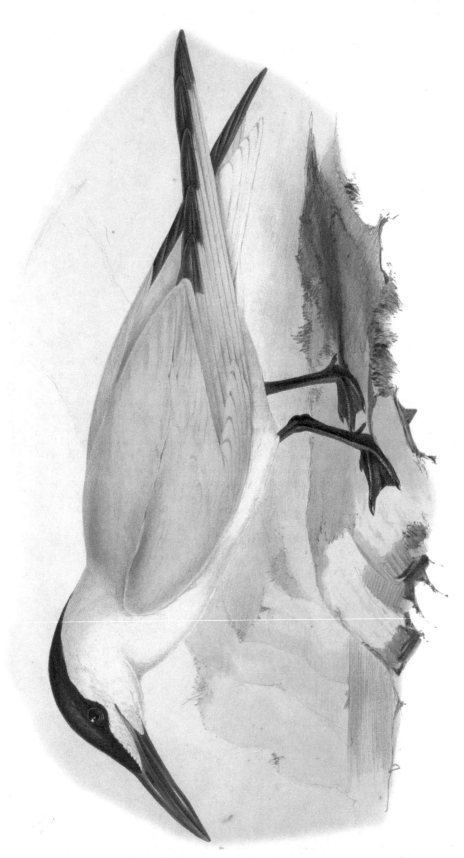

CASPIAN TERN.
Sterna Caspia. (Linn.)

Drawn from Nature & on Stone by J.E Gould.

Printed by C.Hullmandel.

Genus STERNA, *Linn.*

GEN. CHAR.   *Bill* as long or longer than the head, nearly straight, compressed, drawn to a fine point, with both mandibles of equal length, and the upper slightly convex; tomia rather intracted and sharp-edged; lower mandible having a prominent angle near its middle part.   *Nostrils* basal, lateral, linear, oblong, pervious.   *Wings* very long, acuminate, the first quill-feather the longest.   *Tail* more or less forked.   *Legs* having the tibiæ naked for a short space above the tarsal joint.   *Tarsi* short.   *Feet* of four toes, three before, one behind; the three former united by a membrane more or less scalloped, the hind toe small and free.   *Claws* arched and sharp.

CASPIAN TERN.

Sterna Caspia, *Pall.*

L'Hirondelle de Mer Tschegrava.

THIS large and powerful species is dispersed over the northern shores of Africa, the eastern portion of Asia, and all the temperate parts of Europe, where it appears to evince a partiality to inland seas rather than to the wide ocean, and hence we find it most abundant in the Mediterranean, Black, and Caspian Seas, from the latter of which it takes its name.   Of its visits to the shores of Great Britain the instances are but few, and at no regular or definite periods.

In size this noble bird is not exceeded by any other member of its race: it is even larger than many of the Gulls, from which tribe the Terns differ much in their structure, and are moreover destined to fill a very different station in the scheme of creation.

Its food consists of fish, crustacea, mollusca, &c.

The sexes of the Caspian Tern offer no external difference in the colouring of the plumage, but the crown of the head, which is white in winter, becomes on the approach of spring of a deep rich and glossy black, which change is common to both sexes.

The nest is merely a hollow scraped in the sand or shingle; the eggs are four in number, and we have ourselves received them from the small shingly islands at the mouth of the Baltic, which, from the numerous specimens we have seen from that locality, we conceive must form one of the stations to which the Caspian Tern resorts in great numbers for the purpose of breeding: it doubtless also breeds on most of the shores of the Black and other seas before mentioned.

In summer the forehead, crown of the head, and occiput, are black; back, scapulars, wing-coverts, and tail pearl grey; quills greyish brown; the remainder of the plumage pure white; bill rich vermilion; legs and feet black.

The young of the year are clouded and transversely barred with marks of brown, much after the manner of the young of the Sandwich and other European Terns.

The Plate represents a male in summer of the natural size.

SANDWICH TERN.
Sterna cantiaca. (Gmel.)

SANDWICH TERN.

Sterna cantiaca, *Gmel.*

L'Hirondelle de Mer caugek.

LIKE most others of its race, the Sandwich Tern visits the British Isles only during the warmer part of the year, breeding along our shores; and in some localities, as the coast of Kent, Essex, and the Farn Islands off Northumberland, being in considerable abundance. As the severity of winter approaches, and drives into deeper water the young crustacea and fishes on which it feeds, it leaves us for more temperate latitudes, where its food is ever accessible. It is one of the largest of our British Terns, and, unlike some of the genus, is seldom or never seen along inland rivers or upon the large European lakes. Its locality is very extensive, there being few coasts in the Old World where it is not found. In manners and general economy, it differs in no respect from its congeners, being equally remarkable for rapid flight and all that activity and address which fit it for passing over the rough billows of the rock-bound sea.

The process of nidification—for nest it makes little or none—takes place on the naked rock, the shingly beach, or other situations close to the edge of the water. The eggs are two or three in number, marbled with brown or black on a whitish ground.

The male and female offer but little difference of plumage, both being remarkable for a jet black head in summer, which becomes mottled in autumn, and wholly white, or nearly so, in winter. The young, on the contrary, display a very different state of colouring, exhibiting on the upper surface a succession of arrow-shaped marks of black on a light grey ground. In this stage it has been called the Striated Tern by Gmelin and Latham.

In one particular the present bird is very remarkable, having a black beak (the tip alone being yellow in the adult), black tarsi and toes, whereas most of the species of this genus are uniform in the rich red with which these parts are deeply tinted.

In the full plumage of summer, the adult has the head and occiput jet black; the upper parts delicate blueish ash; the sides of the head, the throat and under parts pure white; the bill black with a yellow tip, and the tarsi black. In winter the head is white; and in the intermediate season the progress of change goes on through various stages of mingled black and white, the black of the head returning with the spring.

The young of the first autumn resemble the parents in the colour of the beak and tarsi, except that the former is black to the tip; the upper parts are light grey, the head being barred with transverse semilunar marks of black, and the rest of the upper surface with arrow-headed spots of blackish brown, the quills alone being clear; the under surface white.

The Plate represents an adult of its natural size, and a young bird of the first year in the mottled livery.

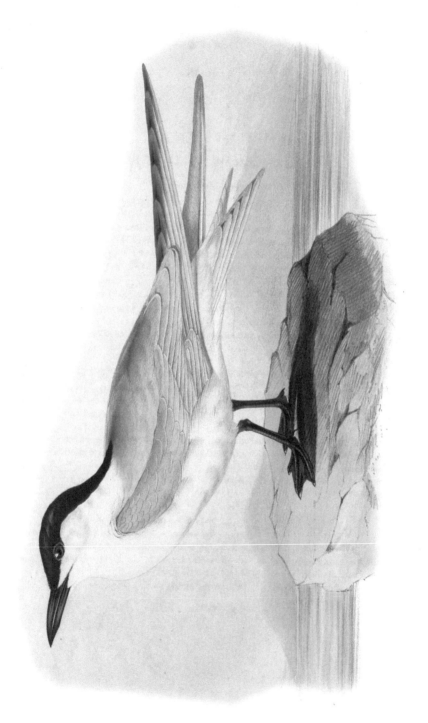

GULL BILLED TERN.
Sterna Anglica. (*Montagu*).

GULL-BILLED TERN.

Sterna Anglica, *Mont.*

Le Hirondelle de Mer Hansel *of Temm.* ?

This rare species of Tern was first made known from specimens obtained in this country by Colonel Montagu, and was described and figured by him in the Supplement to his Ornithological Dictionary. The bill is wholly black, about an inch and a half long, thick, strong, and angulated on the under mandible, at the symphysis or junction of the two portions, in which particular it resembles the Gulls, and this Tern may be considered as a link between the species of the two genera. The upper part of the head, occiput and back of the neck are black in summer; all the upper parts cinereous; outer tail-feathers and all the under parts of the body white; the first five quill-feathers are tipped with greyish black, part of the inner webs white; legs long, exceeding one inch and a half, nearly black; toes long, claws almost straight. In the winter plumage the head is white, with dusky markings about the eyes. Young birds have the head, back and wings mottled with ash colour, light brown and dusky. The sexes are alike in plumage, but the female is rather smaller than the male.

It seems to be now a very general, but not a universal opinion, that the *Sterna Anglica* of Montagu is not the same bird as the *Sterna Anglica* of Temminck's *Manuel d' Ornithologie*, but that this latter bird is identical with the *Sterna aranea* of Wilson and the Marsh Tern of Peale. We have had no opportunity of examining American specimens of this rare Tern, but examples brought from India by Colonel W. H. Sykes were compared, and found to correspond exactly with Colonel Montagu's birds in the British Museum, both in their winter and summer plumage; and that the *Sterna Anglica* of Montagu exists in the Dukhun does not therefore admit of a doubt. Colonel Sykes remarks, that with the aspect, length of wing, lazy flight, and habits of the Tern, this bird has a bill approximating to that of the Gull, and not quite identical with the bill of *Viralva*, under which genus Mr. Stephens has arranged our *Anglica* in his Ornithological portion of Shaw's Zoology, vol. xiii. p. 174.

Numerous fishes were found in the stomachs of the examples of this bird killed in the Dukhun, and this fact is in accordance with the remarks of Charles Lucian Bonaparte, Prince of Musignano, who in his Observations on the Nomenclature of Wilson's Ornithology, states that the habits of the two species of Tern, *Sterna Anglica* and *S. aranea*, are very different; the former is confined to the sea-shore, and feeds sometimes on fishes, while the latter is generally found in marshes, and feeds exclusively on insects.

The Gull-billed Tern is said to frequent, and even to be common on the eastern parts of the European continent, particularly during the summer, where it lays three or four oval-shaped olive-brown eggs, spotted with two shades of darker brown.

We have figured a bird in the summer plumage and of the natural size.

COMMON TERN.
Sterna Hirundo. *(Linn.)*

COMMON TERN.

Sterna Hirundo, *Linn.*

La Hirondelle-de-mer Pierre Garin.

ALL the members of this interesting tribe inhabiting the British Islands are strictly migratory : several species visit us for the purpose of breeding, while others, being inhabitants of more distant countries, are of more rare occurrence.

The Common Tern, although not universally dispersed over our coasts, is nevertheless a very abundant species, being found in great numbers over the southern shores, but more sparingly over the northern, which are almost exclusively inhabited by its near ally the Arctic Tern.

It is now satisfactorily ascertained that the Common Tern does not extend its range to the American Continent, and that its place is there supplied by another species, to which the Prince of Musignano has given the specific appellation of *Wilsonii*, in honour of the celebrated ornithologist by whom it was first described.

How far the Common Tern is distributed over the Old Continent we have not satisfactorily ascertained, but we believe its range is extended from the Arctic Circle to the Mediterranean, and even to the coast of Africa and India, to which southern and eastern countries it is supposed to retire during our winters.

The Common Tern does not confine itself entirely to the sea, but frequently resorts to inland streams, &c. ; and when thus ascending our creeks and rivers these little fairies of the ocean fearlessly fish around our boats, nothing can be more pleasing than to observe their poise and dip. When with their scrutinizing eyes they have observed a fish sufficiently near the surface, they precipitate themselves upon it with an unerring certainty, and a rapidity that is truly astonishing ; this mode of capture strongly reminds us of the *Fissirostral* tribe among the land birds, and they may, indeed, be truly termed the Swallows of the ocean, their long and pointed wings, and small but muscular bodies, being admirably adapted for rapid and sustained flight, and affording the means by which they are enabled to traverse the surface of the deep with never-tiring wings.

The Common Tern breeds upon the sand or shingle beyond high-water mark, making no nest, but scraping a slight hollow for its eggs, which are two or three in number, and which vary much in colour, some being of a deep olive green, while others are of a cream colour, but all blotched with blackish brown and ash grey. " In warm and clear weather," says Mr. Selby, " this bird incubates little during the day, the influence of the sun upon the eggs being sufficient ; but it sits upon them in the night, and also through the day under a less favourable state of the wĕather. The young when excluded are assiduously attended by the parents, and are well supplied with food until they are able to fly and accompany them to sea. During the time of incubation the old birds display great anxiety, and are very clamorous when any one approaches their station, in flying round and frequently descending so near as to strike the hat of the intruder."

Forehead, crown, and occiput black ; back, wings, and tail pearl grey, the edge of the external quill in the wing being black for three parts of its length ; face, sides of the head, neck, and all the under surface white ; bill bright red for two thirds of its length, the tip black ; legs and feet bright red.

The Plate represents two adults, one in the winter and the other in the summer plumage, of the natural size.

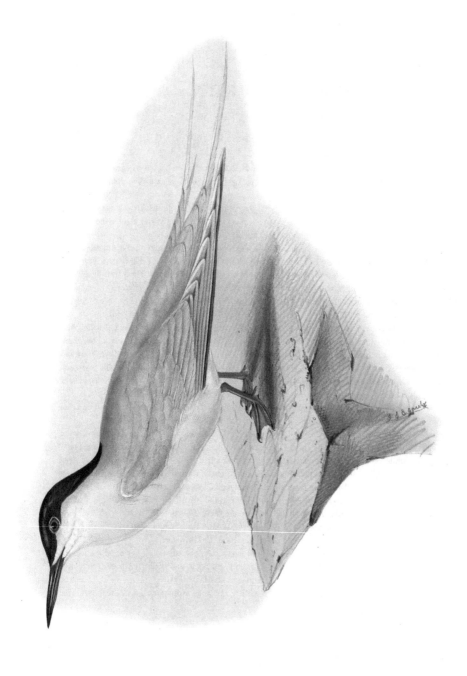

ROSEATE TERN.
Sterna Dougallii. (Mont.)

Drawn from Nature & on Stone by J.& E. Gould.

Printed by C. Hullmandel.

ROSEATE TERN.

Sterna Dougalli, *Mont.*

Le Hirondelle-de-Mer Dougall.

THE delicate rose-colour which pervades the breast of this bird, together with the slender black bill, at once distinguishes it from every species of British Tern; and although not one of the rarest, it is nevertheless more thinly distributed than any other species known to make the British Islands a place of incubation. Continental writers have asserted its occurrence on various northern shores; but it is there also, we have reason to believe, extremely limited in numbers. The first recognition of it as a distinct species is due to Dr. MacDougall of Glasgow, who discovered it breeding on the Cumbray Islands, in the Frith of Clyde. It has since been observed in several other parts of the northern portions of England and Scotland; and it is, perhaps, more abundant in the Fern Islands than on any other part of the coast of this country. Mr. Thompson of Belfast has recently discovered it to be a periodical visitor of the northern coast of Ireland, where it appears to resort annually for the purpose of breeding.

We have ourselves received it rather abundantly from India, particularly the coast of Malabar, a circumstance which is remarkable when we consider that in our latitudes it prefers the more northern parts, being seldom or never seen on the southern coasts of England. From America we believe no examples have yet been seen, nor have we ever observed it from the arctic regions; and as our examples from India were in their full breeding plumage, we are inclined to believe that it is a species which abounds more particularly in the southern regions of the Old World. Mr. Selby, who has seen it in a state of nature, informs us that it is easily to be distinguished while on wing from all other species, its flight being peculiarly buoyant, and sustained by a slower stroke of the pinions: the length of the tail is also characteristic, and its cry is different in expression, resembling the word 'crake', uttered in a tone not unlike that of the Landrail. In the Fern Islands it breeds on the outskirts of the stations occupied by the Arctic Tern, and its eggs much resemble those of that bird, but are a little larger, more pointed at the small end, with the ground colour inclining to cream white or pale wood brown. In habits and manners it scarcely differs from its allied congeners, and it preys on the same kind of fish. The time of its arrival may be stated to be the same as that of the Sandwich and Arctic Terns, and by the end of September nearly the whole of them have departed for warmer latitudes.

The male and female offer little or no difference in the colours of their plumage.

The top of the head and occiput jet black, from the base of the upper mandible; cheeks, throat, and under parts white, delicately tinged with rosy red; the whole of the upper surface delicate grey, with the exception of the outer edge of the first quill-feather, which is black; bill slender, red at the base, the remainder black; tarsi and membrane vermilion.

The Plate represents an adult male in its breeding plumage, of the natural size.

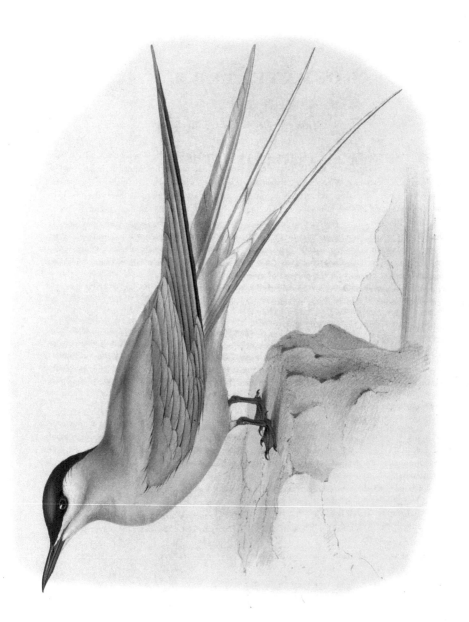

ARCTIC TERN.
Sterna Arctica, *(Temm.)*

ARCTIC TERN.

Sterna Arctica, *Temm.*

L'Hirondelle de Mer arctique.

It is to M. Temminck that we owe the knowledge of the present bird as constituting a different species from that of the Common Tern (*Sterna Hirundo*, Linn.), to which it bears so close a resemblance as almost to require actual comparative examination of the two species, to determine the characters which form the line of distinction:—the accurate representation, however, which we have given of both species, with the minute indications pointed out in the letter-press, will, we trust, clear up every difficulty attached to these two species, so nearly allied, and so often confounded. We have ourselves had abundant proofs that the present bird is a constant inhabitant, in considerable numbers, of many parts of our coast, but more especially its northern portion, and the adjacent Islands the Orkneys and Shetland, where it is known to breed regularly; and it is not a little singular, according to the most credible information, that these Terns, although bearing so close an affinity to each other, do not associate together at the same breeding-places, but that each retains its peculiar locality although both breed in the immediate neighbourhood of each other. Thus one species will occupy an island, or a portion of it, to the entire exclusion of the other, and *vice versâ*. M. Temminck informs us, that it is especially common in the Arctic circle, which he considers to be its true habitat, and where it occupies the place of the *Sterna Hirundo* of more southern latitudes. We have had opportunities of examining this species in all its stages, and we find that they strictly correspond with those of its allied congeners. The young offer also but little difference from those of our Common Tern. There is, however, one infallible rule by which not only the adult but the young in any stage may be at once discriminated, viz. by a comparison of the length of the beak and tarsus, characters on which the greatest reliance may always be placed. The Arctic Tern is altogether smaller and more slender, with a longer and more elegant tail, the beak wholly red and much less robust, as well as a quarter of an inch shorter, measuring from the gape to the tip; the tarsi are also proportionately smaller, measuring in length only seven lines; to which may be added that its colour is much more uniform, nearly the whole of its body, both above and below, being covered by a blueish ash colour; the head and back of the neck black.

It breeds among the shingles on the sea shore, the female laying two or three eggs very similar in colour and markings to those of the Common Tern, but smaller.

We have figured a male in its summer plumage.

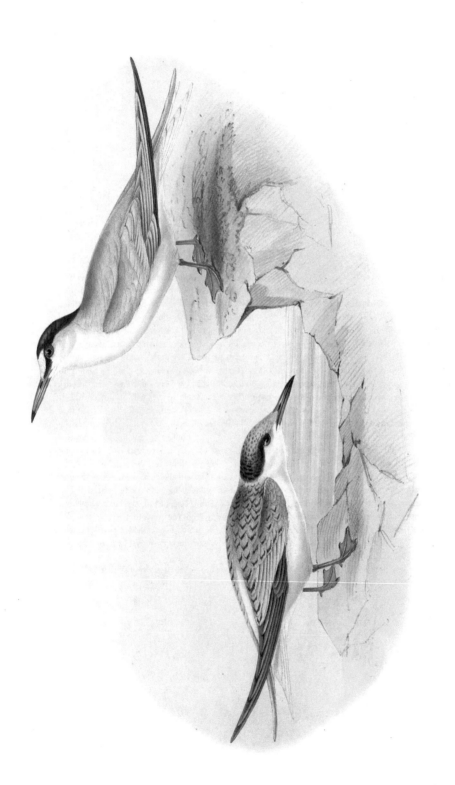

LITTLE TERN.
Sterna minuta. *(Linn.)*

LITTLE TERN.

Sterna minuta, *Linn.*

La Petit Hirondelle de Mer.

THIS elegant species of Tern appears to be more generally distributed than any other of its race : we have ourselves received specimens from various parts of India, which, with others from America, prove to be strictly identical with those found in Europe. It visits the coasts of England in great numbers on the approach of spring ; and not our coasts alone, but those of the Continent to a considerable extent northwards, apparently arriving from more southern latitudes where it has been passing the winter. It associates in large flocks, and may be observed winging its way over the surface of the sea, particularly off coasts which are flat, low, and shingly, and where small islands are left by the retiring tide. These islands afford for them and many other maritime species a resting-place, where they collect in considerable numbers until the rising waters force them to take wing again. The adjoining shingly beach also affords them a place of nidification, their eggs being deposited in a slight depression among the broken shells just above high-water mark ; and so closely does the colouring of the eggs assimilate with the mingled mass around them, that they escape the casual search of inexperienced eyes.

Winged insects, small fishes, and marine animals, form their food : these latter they take from the surface of the water as they fly, but without diving, a power which the Terns, as we scarcely need observe, do not possess.

The young and adult offer marked differences of colouring. The former, before the autumn moult, have the forehead yellowish white ; the top of the head, occiput, and back of the neck, brown with black bars ; a black stripe passes through the eyes ; the back and wings light brownish grey, each feather having a border of blackish ; tail- and quill-feathers tipped with whitish. After the autumn moult the back of the head is black, and the under parts are, as in the adult, light grey.

The adult plumage may be thus characterized. The forehead white ; a black stripe passes from the base of the upper mandible, through the eye, and joins a large black patch on the occiput and back of the neck ; upper parts fine blueish grey, the rump and tail being white, as are also the under parts ; beak orange tipped with black ; tarsi orange ; length 8¼ inches, but when seen with outspread wings flapping over the sea, the bird appears much larger than it is in reality.

Our Plate represents an adult and a young bird of the natural size.

NODDY TERN.
Sterna stolida. *(Linn.)*

Drawn from Nature & on Stone by J.&E. Gould.

Printed by C. Hullmandel

NODDY TERN.

Sterna stolida, *Linn.*

Le Mouette brun, ou Le Fou.

Two individuals of this species having been shot in the summer of 1830, off the coast of Wexford in Ireland, between the Tasker Lighthouse and Dublin Bay, it becomes necessary to include a figure of it in the present work. These examples (the first, we believe, that have occurred in Europe,) have been placed on record by W. Thompson, Esq., Vice-President of the Belfast Natural History Society.

We are indebted to American ornithologists for the best accounts of the habits of the Noddy. Mr. Audubon found numbers collecting from all parts of the Gulf of Mexico and the coasts of Florida, for the purpose of resorting to their breeding-places on one of the Tortugas called Noddy Key, where many of these birds were observed by this gentleman in May 1832 repairing old nests that had been used by themselves or their companions the preceding year.

The Noddy, unlike the generality of the Terns, builds in bushes or low trees, making a large nest of twigs and dry grass, while hovering over or near which the old birds utter a low querulous murmur: the eggs are three in number, of a reddish yellow colour, patched and spotted with dull red and purple. The young birds are said to be excellent eating. Unlike the other members of its family the Noddy takes its prey while skimming along the surface of the water; and the old birds seek their food, which consists principally of small fishes, at a greater distance from land than Terns are generally observed to do.

Mr. Nuttall states in his Manual of the ornithology of the United States and of Canada, that " the Noddies breed in great numbers in the Bahama Islands, laying their eggs on the shelvings of rocks ;" and we observe that Dr. Latham, on the authority of others, makes a similar statement; he was told also that these birds lay their eggs in vast numbers on certain small rocky islands near St. Helena.

Forehead white passing into grey on the back of the head; immediately before the eye a patch of deep black; throat and sides of the face greyish brown; primaries and tail deep blackish brown; the remainder of the plumage dull sooty brown, inclining to chocolate; bill and feet black.

We have figured an adult of the natural size.

BLACK TERN.
Sterna nigra. (*Linn.*)
Viralva nigra. (*Leach*).

Drawn from Life & on Stone by J & E Gould.

Printed by C Hullmandel.

Genus VIRALVA, *Leach*.

GEN. CHAR. *Beak* shorter than the head, subulated, nearly straight, slightly compressed, the tip a little inclined: the *upper mandible* nearly straight. *Nostrils* oblong, basal. *Wings* long. *Tail* slightly forked. *Feet* four-toed, slender; the *hinder toe* minute. *Claws* small.

BLACK TERN.

Sterna nigra, *Linn*.

Viralva nigra, *Leach*.

L'Hirondelle-de-mer epouvantail.

ALTHOUGH we have given the generic characters of the genus *Viralva* of Dr. Leach, comprehending those Terns which have their tails almost square, in addition to other less important characters; we still hesitate to adopt the genus of this naturalist, on the ground, that the separation is established on characters too trivial in our opinion to substantiate a new genus. It will, however, be our aim to give a complete account of the habits of the present bird and the other European species included in this genus, leaving our readers to adopt it, or not, as they think best.

The Black Tern is an abundant species, but confined more especially to the vicinity of large rivers, fresh-water lakes, and morasses, particularly those of Holland and Germany, and extending thence as far as the Arctic circle. It annually visits the fens and marshy districts of this country, such as those of Cambridgeshire, Norfolk, and Lincolnshire, for the purpose of incubation, but is not near so abundant now as formerly. This species assembles to breed in flocks more or less numerous among the flags and willows which border the edges of water, laying four or five eggs of a dark olive brown, marked with numerous spots of darker brown and black.

The Black Tern differs in its habits, nanners, food, mode of nidification, the situations it selects for that purpose, and its manner of flight, from the true Terns, which may at once be distinguished from it by their very long wings, and swallow-like form of tail, and by their giving a preference to the sea and its inlets, where they obtain their food, which consists in a great measure of small fish, mollusca and other marine productions; but in the present bird we find the wing less elongated, and the tail less forked, the tarsi longer, and the toes less webbed, while the food is taken almost solely during flight, and consists of winged insects, such as moths, flies, and the larger species of gnats, to which are added aquatic larvæ, and occasionally small fishes.

The flight of the Black Tern also, instead of that heavy flapping motion which characterizes the oceanic Terns, is smooth and rapid, while the bird continues to pass and repass over the same space like the Swallow in search of its insect food. In England, the Black Tern appears to be migratory, leaving us after the breeding season is over, and returning the following spring; a circumstance we should consider to take place also in the northern portions of Europe. Although the young of all the Terns differ in colour from the adult bird, the contrast of the present species is the most remarkable, the colouring of the two being almost diametrically opposite; they notwithstanding soon assume the adult state of colouring, and in about eight months gain their mature livery. The males and females are alike in plumage.

The adult birds have the beak, head, neck and breast black, becoming paler on the abdomen; the whole of the upper surface and tail of a fuliginous grey; vent and under tail-coverts white; legs dusky red; irides brown. The young have the forehead, cheeks, neck, and whole of the under surface of a pure white; the top of the head and occiput greyish brown; the back, wings and tail grey, intermingled with brown.

We have figured an adult and a young bird of the natural size.

WHITE-WINGED TERN.
Sterna leucoptera. *(Linn.)*
Viralva leucoptera. *(Leach.)*

Drawn from Nature & on Stone by J & E Gould.

Printed by C. Hullmandel.

WHITE-WINGED TERN.

Sterna leucoptera, *Linn.*

Viralva leucoptera, *Leach.*

L'Hirondelle-de-mer leucoptère.

THE White-winged Tern would appear to represent in the southern districts of Europe the Common Black Tern of the more northern latitudes, frequenting, like that bird, inland lakes and marshes, as well as the low flat borders of the sea. We are informed that it inhabits all the bays and gulfs along the shores of the Mediterranean, and that it is very common in the neighbourhood of Gibraltar : it visits also, according to M. Temminck, the lakes and marshes of Italy, such as Lucarno, Lugano, Como, &c., but never extends its journey to Holland or the parallel latitudes.

In habits, manners, size, and structure, it strictly resembles the Black Tern ; the pure whiteness of its tail, and the greyish white of its wing, will, however, serve at once to distinguish it from that species.

Its food consists of insects, particularly dragon flies, moths, and other winged and aquatic insects, worms, and occasionally small fishes.

Of its nidification and the number and colour of its eggs, little is at present correctly ascertained ; but we have every reason to believe that they bear a close resemblance to those of the Black Tern.

The sexes do not differ in the colour of their plumage, but the young of the year have less white on the wings, and the rest of the plumage is of a lighter and browner hue ; in fact, it undergoes a change very similar to that of the Black Tern.

The whole of the head, neck, back and belly, and the two outer quill-feathers black ; the remainder of the wings greyish white ; the rump, tail, vent, and under tail-coverts white ; beak brownish red ; tarsi brownish red.

The Plate represents an adult of the natural size.

MOUSTACHE TERN.
Sterna leucopareia. *(Natt.)*
Viralva leucopareia. *(Steph.)*

Drawn from Nature & on Stone by J&E. Gould.

Printed by C. Hullmandel.

MOUSTACHE TERN.

Sterna leucopareïa, *Natt.*

Viralva leucopareïa, *Steph.*

L'Hirondelle-de-mer moustac.

FOR the knowledge of this species we are indebted to M. J. Natterer of Vienna, who discovered it in the marshes of Hungary. It has also been seen by M. Temminck in Capo d'Istria on the coast of Dalmatia.

The Moustache Tern, like its black- and white-winged brethren, appears to prefer inland and extensive marshes rather than the ocean. Like the last-mentioned species it is almost confined to the eastern portions of the Continent. In Europe it is perhaps one of the rarest of its tribe, and although its habits are but little known, we may reasonably conclude from its peculiar form that its general economy is strictly similar to the other Viralves, or Marsh Terns.

Its food consists of the winged insects inhabiting the marshes, to which are added worms, snails, &c.

The sexes offer no perceptible difference in their plumage. The young, says M. Temminck, have the top of the head of a reddish colour varied with brown; the occiput, the region behind the eyes, and the orifice of the ears of a blackish ash; the back, scapularies, secondaries, and quills brown in the middle, bordered and terminated with yellowish brown; the tail-feathers blackish ash with the exception of the outer ones, which are tipped with white; beak brown, reddish at the base; feet flesh colour.

According to M. Temminck's description the adults are subject to considerable seasonal changes; for he informs us that in winter the top of the head, the occiput, and all the under parts are pure white; a black spot is situated behind the eye; the back, rump, tail, and wings are of a clouded ash grey; the beak and feet deep lake red; the irides black; and that in spring the whole crown of the head is deep black, and the under surface clouded with blackish ash, becoming lighter towards the throat, and leaving the sides of the face and ear-coverts pure white, whence it has received the appellation of moustache.

The Plate represents a male of the natural size in its full spring dress.

LAUGHING GULL.
Xema ridibundus: (Boye)

Genus XEMA.

GEN. CHAR. *Beak* short, slender, straight, laterally compressed, its tip bent down; the lower mandible somewhat angulated beneath. *Nostrils* very slender, linear. *Legs* slender; *tibiæ* naked on the lower part. *Tail* forked.

LAUGHING GULL.

Xema ridibundus, *Boje.*

La Mouette rieuse ou à Capuchin brun.

THE characters which distinguish the genus *Xema* of Dr. Leach from the genus *Larus*, consist not only in a decided difference of form, but in certain points of colouring, and the changes which the species comprised in it undergo at different seasons; for example, the bill and legs are bright red, and the head changes in spring from white to black or deep chocolate brown, which latter colouring is certainly confined to the breeding-season, and disappears on the approach of autumn; in addition to this we find that the young pass through a very different gradation of plumage to that which obtains among the Gulls in general. Independently of these variations in the colouring, we may observe that the general contour of the species is much more light and elegant, the bill more feeble, and the tarsi more slender; they choose, moreover, a very different place for the purposes of nidification, always resorting to low flat lands, often some distance from the sea, the nest being placed on the ground, whereas the generality of the Gulls build upon ledges of rock bordering the sea.

Of all the species comprised in the present group which inhabit our island, the Laughing Gull is by far the most common and perhaps the most elegant of its genus. During the summer it resorts in immense flocks, for the purpose of nidification, to many of our marshy islands near the coast, after which it again returns to the sea, or the mouths of large rivers, and is found at this season round the whole of our coasts, but is not then to be distinguished by the bright chocolate colouring of the head, which character is so remarkable during the breeding-season. In general habits, manners, and mode of flight, it agrees with the rest of the Gulls; though, as its light form and long tarsi sufficiently indicate, its actions on the ground are much more nimble and rapid. It is said to be a bird of passage in Germany and France, but is found in the greatest abundance in Holland throughout every season of the year. Its food consists of various insects, worms, mollusca, and small fishes.

In its full summer plumage the bill, naked skin round the eye, and tarsi, are bright red; the whole of the head and throat deep chocolate brown; the back and shoulders delicate grey; quills white on their outer edges, with the exception of the first, in which it is black, the extremities of all the rest being black slightly tipped with white; rump, tail, and whole of the under surface white.

The winter plumage is similar to that of summer, with the exception of the chocolate hood, which is gradually exchanged for pure white, a change which Mr. Yarrell has correctly observed in his valuable paper "On the Laws which appear to influence the assumption and changes of plumage in Birds," published in the Transactions of the Zoological Society of London, (vol. 1. part 1. p. 13,) is produced not by a process of moulting, but by an alteration in the colour of the feathers.

The young of the year have the colour of the bill and tarsi much more obscure; the top of the head and ear-coverts are mottled with brown, which is also the colour of the back and shoulders, each feather having a lighter margin; the tail is broadly edged with black.

The full plumage of maturity is not acquired until after the moulting of their second autumn, and is assumed by gradations. The sexes do not differ in their colouring.

The Plate represents an adult, and a young bird of the year, of the natural size.

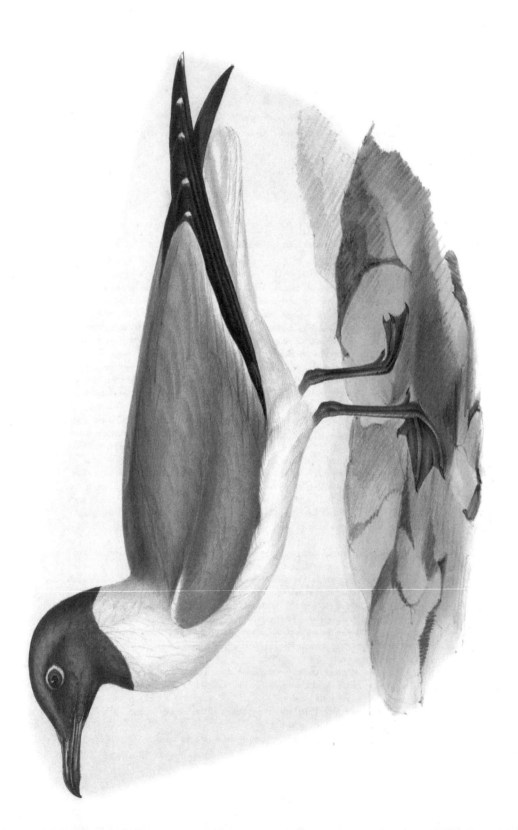

BLACK-WINGED GULL.
Larus atricilla. (Linn.)

BLACK-WINGED GULL.

Xema atricilla,

Larus atricilla, *Linn.*

La Moutte à ailes noires.

In figuring this species of Gull under the specific title of *atricilla*, we would beg to observe that it should not be confounded with the *atricilla* of M. Temminck, which name must necessarily fall in consequence of its having been previously given to another species.

The present bird is common in the United States of America, and was, we believe, the only species of Gull figured by Wilson, who considered it to be the true *atricilla* of Linnæus. Of the capture of this bird in Europe no later account has been published than that given in the publications of Montagu, whose original specimen, now in the British Museum, has afforded us an opportunity of determining it to be identical with the American bird. Beneath we have annexed the account given by Montagu, who clearly points out the distinctions between it and the common species, *Xema ridibundus*. "In the month of August 1774, we saw five of them feeding in a pool upon the shingly flats near Winchelsea; two only were black on the head, the others were mottled all over with brown. One of them was shot, but although the remaining four continued to resort to the same place for some time, the old ones were too shy to be procured. We also saw two others near Hastings in Sussex. They may easily be known from the Black-headed Gull even while flying; the flight is different, the bird appears much larger and the tail shorter in proportion."

In its habits, manners, mode of nidification and food, this species closely resembles its congeners, feeding upon the refuse of the fisherman or any animal substances thrown up by the tide: it may also be observed scattered over marshes and newly-ploughed fields busily engaged in searching for worms, insects, and their larvæ. It usually breeds in the marshes near the coast, and lays three eggs, of a dull clay colour, thinly marked with irregular patches of pale purplish brown.

The whole of the head and the upper part of the neck are blackish lead colour; circle surrounding the eye, the neck, all the under surface, and tail pure white; primaries black, with the exception of their extreme tips, which are white; the remainder of the upper surface and wings dark grey passing into white on the edge of the shoulder and the tips of the secondaries; bill red; feet reddish brown.

We have figured an adult of the natural size.

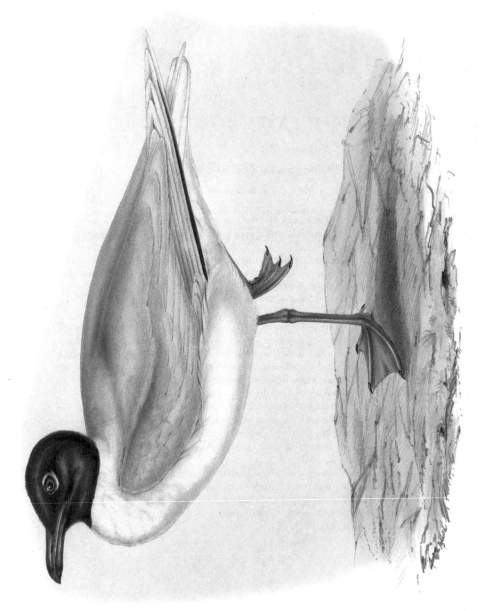

BLACK HEADED GULL.
Xema melanocephalus, (Boje).

BLACK-HEADED GULL.

Xema melanocephala, *Boje*.

Le Mouette à capuchon noir.

This species of *Xema* may be readily distinguished from its European congeners by the jet black colouring of the head, the robust bill, and the silvery whiteness of the tips of the primaries. Several examples of this fine species in different stages of plumage were obligingly sent us by M. Temminck, from one of which the accompanying representation of the bird in its full summer dress was taken. In winter, the conspicuous black hood is entirely changed to white, as is the case with all the other species of this remarkable genus.

The native habitat of the Black-headed Gull is the more southern parts of Europe, particularly along the shores of the Adriatic. It is said to be very abundant in Dalmatia, inhabiting and breeding in the marshes in the same manner as the Laughing Gull (*Larus ridibundus*, Linn.), so common in our island.

Its food consists of small fishes, snails, and various species of marine insects.

The whole of the head, with the exception of a small patch of white above and below the eye, is of a profound black; the back of the neck, chest, tail, and all the under parts pure white; the whole of the upper surface delicate pearl grey; the primaries white at their tips; the outer web of the first primary is black for three parts of its length from the base; beak, legs, and eyelids vermilion; irides brown.

The figure is of the natural size.

LITTLE GULL.
Xema minutus. *(Boje)*.

Drawn from Nature & on Stone by J. & E. Gould.

Printed by C. Hullmandel.

LITTLE GULL.

Xema minutus, *Boje.*

Le Mouette pygmée.

OF all the Gulls which frequent the British shores, the present species is by far the smallest as well as the rarest: nor is it much more common in the western part of the European continent, its native habitat being more especially the eastern portions of Russia, Livonia, Hungary, and the shores of the Black and Caspian Seas; and although it has been considered as identical with a species common to America, we have the best reasons for affirming it to be entirely distinct. It was first noticed as occurring in our island by Colonel Montagu, since which period examples have been killed at different times, and we have ourselves had the pleasure of receiving it in a recent state.

Like the other species of the present genus, the Little Gull is a bird which exhibits a remarkable disparity of colouring in the winter and summer states of plumage, as well as from youth to maturity: all the examples killed in our island have been either in their immature or winter plumage. In summer it is characterized by a black head, which colouring it loses before the approach of winter; but in all the winter-killed specimens we have had an opportunity of examining, traces of this summer plumage remained, and in this state we have figured an adult, which is represented by the foremost bird in the Plate. Its flight is as light and buoyant as can well be imagined, and its general actions and form resemble those of the rest of the genus.

The colouring of the adults in their winter plumage is as follows:

The whole of the upper surface is of a beautiful bluish ash; the quills and secondaries tipped with white; the throat and under-surface pure white, with a slight tinge of rose-colour; bill brownish red; tarsi bright red; irides brown.

In summer the whole of the head and the upper part of the neck become of a brownish black.

The young when a year old resemble the adult in the winter plumage, with this exception, that the shoulders, scapulars, quill-feathers, and tip of the tail are deep brownish black, and that the beak and legs are not so red.

The Plate represents an adult in its winter plumage, and a young bird of the first year, of the natural size.

SABINE'S GULL.
Xema Sabini. *(Leach)*

Drawn from Nature & on stone by J. & E. Gould.

Printed by C. Hullmandel.

SABINE'S GULL.

Xema Sabini, *Leach.*

La Mouette de Sabine.

This species has been added to the British Fauna in consequence of two examples having been killed in Belfast Bay and one in Dublin Bay, of which notices have been recorded in the 5th No. of the Magazine of Zoology and Botany; it has therefore become necessary to include a figure of it in the present work. It is almost strictly an arctic species, and as we have nothing to add to the account of its natural history published by Dr. Richardson, we prefer quoting the words of this scientific traveller as given in the Fauna Boreali-Americana, to recording the same facts in any language of our own.

" This interesting species of Gull," says Dr. Richardson, " was discovered by Captain Edward Sabine. It was first seen on the 25th of July, at its breeding-station on some low rocky islands, lying off the west coast of Greenland associated in considerable numbers with the Arctic Tern, the nests of both birds being intermingled. It is analogous to the Tern not only in its forked tail, and in its choice of a breeding-place, but also in the boldness which it displays in the protection of its young. The parent birds flew with impetuosity towards persons approaching their nests, and when one was killed, its mate, though frequently fired at, continued on the wing close to the spot. They were observed to get their food on the sea-beach, standing near the water's edge, and picking up the marine insects which were cast on shore. A solitary individual was seen in Prince Regent's Inlet, on Sir Edward Parry's first voyage, and many specimens were procured in the course of the second voyage on Melville Peninsula. Captain Sabine also killed a pair at Spitzbergen, so that it is a pretty general summer visiter to the Arctic seas, and is entitled to be enumerated amongst the European as well as American birds. It arrives in the high northern latitudes in June, and retires to the southward in August. When newly killed it has a delicate pink blush on the under plumage. The eggs, two in number, are deposited on the bare ground, and are hatched in the last week of July. They are an inch and a half in length, of an olive colour with many dark brown blotches."

In summer, the head and upper part of the throat are blackish grey, bounded below by a collar of velvet black; the mantle and wings bluish grey; greater coverts and primaries deep black, the latter tipped with white; edge of the shoulder and the extremities of the secondaries white, forming an oblique band across the wing; neck, all the under surface, and tail pure white; bill black at the base, and yellow at the tip; eyelids red; irides, legs, and feet black.

The young birds of the year have the head mottled with blackish grey and white; back, scapulars, and wing-coverts blackish grey tinged with yellow brown; wing-primaries white with black ends; throat and breast pale ash colour; belly white; upper and under tail-coverts white; tail-feathers white tipped with black.

We have figured an adult male in the summer plumage, of the natural size.

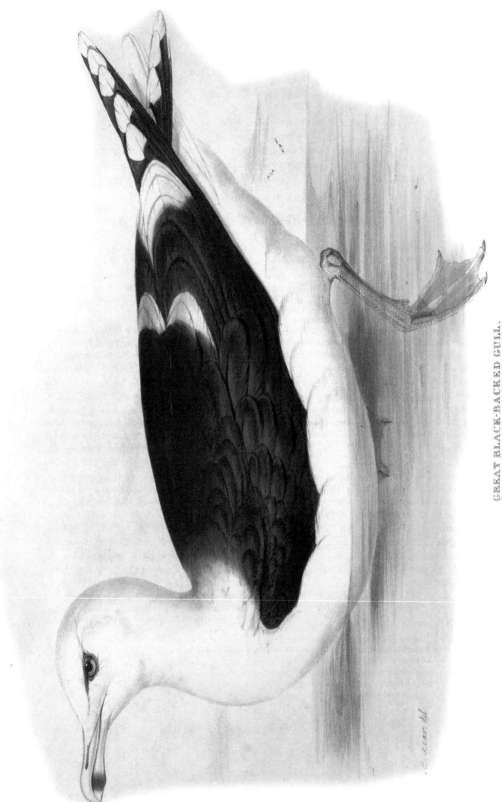

GREAT BLACK-BACKED GULL.
Larus marinus. (Linn.)

Genus LARUS.

GEN. CHAR. *Bill* of mean length, strong, straight, cultrated, the upper mandible having the tip incurved; symphysis of the upper mandible strongly angulated, and ascending from thence to the point. *Nostrils* placed in the middle of the bill, lateral, oblong, narrow, and pervious. *Tongue* pointed, with the extreme tip cloven. *Wings* long, acuminated. *Tail* even, or slightly forked. *Legs* placed near the centre of the body, of mean length and strength, with the lower part of the tibiæ naked. *Feet* of four toes, three before and one behind; the three in front united by a membrane; the hind one short and free.

GREAT BLACK-BACKED GULL.

Larus marinus, *Linn.*

La Goêland à manteau noir.

THIS fine species of Gull, of which an adult in its winter plumage is represented in the accompanying plate, is rather abundantly dispersed round the shores of our island as well as on the opposite coasts of France and Holland. Three years at least are required to accomplish the plumage of maturity; hence by far the greater number of those which are captured are yet in their youthful dress, which differs so much from th a of the adult as to have caused considerable confusion in its nomenclature. Even the large surface of our publication will not admit of our illustrating the present species of the natural size, and we have not deemed it necessary to insert a figure of the bird in its youthful state; this deficiency will, however, be remedied by our figures of the old and young of the Lesser Black-backed Gull, which resembles the species in question in every respect except in size, and which undergoes precisely the same changes.

Many authors have asserted that the *Larus marinus* is not an inhabitant of America, while others have stated that it is there a bird of considerable rarity. Mr. Audubon has, however, just sent forth to the world a magnificent drawing of an individual shot by himself within the United States; thus satisfactorily proving that the New World is included in its range. It is widely distributed along the shores of the European Continent, more particularly the seas of its northern regions. The British Islands afford several localities which are resorted to by this Gull for the purpose of breeding, among which, according to Mr. Selby, may be enumerated the steep holmes and sandy islands in the Bristol Channel, Souliskerry in the Orkneys, the Bass Island in the Frith of Forth, and one or two stations on the Scottish coast.

It breeds also in the marshes at the mouth of the Thames, making a nest on the ground, of reeds, rushes, and flag leaves. The eggs are three in number, like those of the Herring Gull in shape, but larger; the ground colour of various shades of brown, always blotched and spotted with darker brown.

On the water it is extremely light and buoyant, swimming with little exertion, gracefully rising and falling with the undulating waves of the ocean; and, being capable of sustaining a long and continued flight, constantly wandering over the surface of the water or searching along the coast with every rising and receding tide, it seldom fails to find for the gratification of its omnivorous appetite a plentiful supply of half-decomposed animal substances, the refuse from ships, marine crustacea, &c.

The female differs from the male only in being rather smaller, and in her colouring being somewhat less intense.

The only change of plumage which this Gull undergoes after having attained maturity is in the snowy white head and neck of summer giving place to a mottled grey colouring of these parts in winter.

The young of the first and second year is distinguished by a mottled grey and white plumage covering the whole body. In this stage of its existence both this and the young of the Lesser Black-backed Gull have been known under the names of Wagel, Cobb, &c.

The adult has the head, neck, throat, tail, and all the under surface pure white; the mantle and wing-coverts greyish black; the primaries and secondaries black tipped with white; legs pale pinkish white inclining in some individuals to fleshy red; bill pale yellow with a blood-red spot on the angular projection of the lower mandible; irides yellowish hazel.

The figure is about two thirds of the natural size.

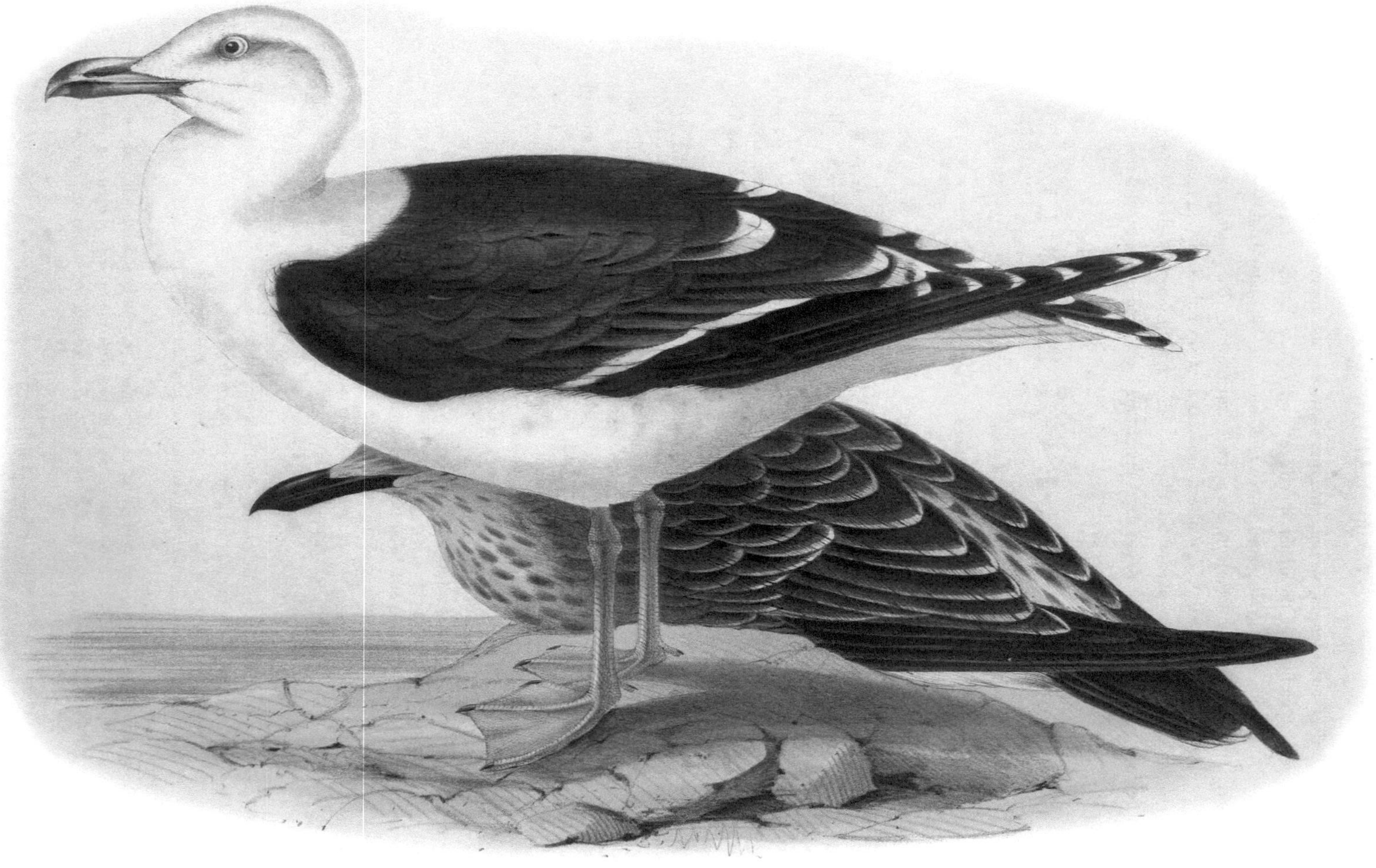

LESSER BLACK-BACKED GULL.
Larus fuscus; *(Linn)*

LESSER BLACK-BACKED GULL.

Larus fuscus, *Linn.*

La Goëland à pieds jaune.

THIS species so nearly resembles the *Larus marinus*, or Great Black-backed Gull, that were it not for the different colour of its tarsi, and their greater length in comparison to the size of the bird, its inferiority in size would be the only distinguishing difference : and indeed, except in these particulars, so closely do they resemble each other, that on a casual view the two species might readily be thought identical or mere varieties : the above-mentioned characters being, however, permanent, no doubt can possibly arise as to their being really distinct.   Not only do they closely resemble each other in their form and colouring, but they are also very similar in their habits and manners, both species breeding alike on our shores and both being permanent residents on the British Islands, particularly on the Northumbrian shores and in some districts of Scotland, where, as Mr. Selby informs us, it may be found at all seasons of the year ; he also adds, that it breeds abundantly on the Fern Islands.

The Lesser Black-backed Gull, inhabiting as it does the borders of the sea, depends for its subsistence, like the rest of the genus, upon the produce of that element, feeding upon fish, mollusca, &c., in search of which it sails to and fro at no great distance from land : it is also observed to frequent pastures or newly ploughed fields near its usual resort in search of worms, larvæ, and insects.   It wanders far up the mouths of large rivers, and in winter is occasionally seen upon the larger inland lakes.   On the shores of the continent of Europe its habitat is spread from the Baltic to the Mediterranean.

It builds in morasses and on the rocks near the sea-shore, in which particular alone it differs from the larger species, the nest being formed of dried grass ; the eggs are three or four in number, of a deep olive green irregularly blotched with brownish black.

As is the case with the whole of the tribe to which it belongs, the young and old offer a marked contrast in their colouring ; the youthful dress being characterized by a plumage of mottled white and brown, which is not wholly lost until the third year ; in this state it has not only been confounded with the young of the Great Black-backed Gull, but has also been considered by many as a distinct species ; hence Brisson described it under the name of *la Mouette grise*, in which opinion Storr coincided.

The sexes are alike in plumage, but in winter the head and neck of the adult have every feather streaked down its middle with a dash of brown, which disappears in summer, leaving those parts of a pure white : the back, which does not alter, is of a fine bluish black : the quill-feathers are black ; the two external ones have a white oval mark near their tips, the remainder tipped with white : the whole of the head, neck, rump, tail and under surface pure white : beak fine light yellow, with the exception of the angle, which is red : irides light yellow : tarsi yellow.

The young of the year have the throat and the fore part of the neck greyish white streaked longitudinally with brown ; the neck and under surface white, largely blotched with deep brown ; the feathers of the upper surface blackish brown on the middle, with a lighter margin ; quills deep black with a long white mark towards the tip ; the tail grey at the base, the rest being black in the centre and fading off to white at the tip ; beak black at the point and brown at the base : tarsi dull yellow : irides brown.

The Plate represents an adult and a young bird of the year, rather more than two thirds of the natural size.

GLAUCOUS GULL.
Larus glaucus. (Brunn.)

Edwd Lear del.

Printed by C. Hullmandel.

GLAUCOUS GULL.

Larus glaucus, *Brunn.*

La Goëland Burgermeister.

THIS noble species fully equals in size, if it do not exceed, the Great Black-backed Gull (*Larus marinus*, Linn.), but from which it may at all times be distinguished by the extremely delicate grey colouring of its mantle and upper surface. Although frequently occurring on our coast, the individuals taken are in the proportion of about twenty young birds to one adult, immature birds having a more decided propensity to wander far from their native habitat than the old ones.

As its pale and almost white colouring would seem to indicate, the Glaucous Gull is a native of the high polar regions, where it frequents shores bound up by ice and snow, crags of ice, and floating icebergs. The British Islands lying, as it were, directly in its way during its wanderings southward, is one reason why this Gull is more abundant with us than it is on most of the other coasts of Europe; but, as we have stated above, these visitants are principally young birds.

In its habits and manners it perfectly agrees with the rest of the larger Gulls; if anything, its flight is more buoyant and easy, which may be occasioned by its denser plumage.

It breeds on precipitous rocks, and its eggs are stated to be of a pale purplish grey with spots of umber brown.

It is a bird of voracious appetite, preying not only upon fish but upon every kind of carrion; nor are small sea birds free from its attacks, and there is some reason to suppose that the Little Auk frequently falls a prey to its craving appetite. Although the rocky shores of Norway and Sweden as well as those of Holland and France are visited by this Gull, they do not come within the number of its breeding-places.

The adult bird in summer has the whole of the plumage pure silky white, with the exception of the mantle and wing-coverts, which are delicate grey; bill pale yellow, with the angular projection of the lower mandible blood red; legs and feet flesh colour. In winter the head and sides of the neck are streaked with pale brown, which disappears on the approach of spring.

The young birds have the whole of the plumage of a dirty greyish white, spotted and barred all over with greyish brown; the shafts of the primaries white; the bill reddish yellow at the base and black at the tip; and the feet pale flesh colour.

Our Plate represents an adult male in the summer plumage and a young bird about two thirds of the natural size.

ICELAND GULL.
Larus Islandicus. *(Kilnewsten.)*

Drawn from Nature & on stone by J & E Gould.

Printed by C. Hullmandel.

ICELAND GULL.

Larus Islandicus, *Edmonston.*

Larus glaucoïdes, *Temm.*

La Mouette d'Icelande.

THE occurrence of this beautiful species of Gull on our coast is more frequent than is generally supposed; but it appears to have escaped observation in consequence of its close resemblance to the *Larus glaucus*, which resemblance is even more striking in the young birds, and it is seldom that any but immature birds of either species are captured in any of the temperate portions of Europe.

The northern regions constitute the native habitat of this Gull, whence it is driven southward by the extreme severity of the weather; the young, as is usually the case in migratory birds, wandering to the greatest distance from home. Considerable confusion, it would appear, has existed respecting the nomenclature of this species, but this has been so carefully cleared up by Mr. Selby, that we have taken the liberty of quoting in full his observations. " In Mr. Edmonston's first notice of the *Glaucous Gull*, under the name of *Larus Islandicus*, a suspicion is started, from the difference of size existing between individuals of the newly observed kind, that there might be two species, having such a relation to each other as that between the Greater and Lesser Black-backed Gulls (*L. marinus* and *L. fuscus*). This upon further investigation was found to be actually the case; and some interesting remarks upon the new species, by the same gentleman, were afterwards published in the latter part of the fourth volume of the Wernerian Society's Memoirs, where he has appropriated to it the specific title of *Islandicus*, having then ascertained that the larger species previously noticed, and to which he had applied the term, was already recorded, and generally known by the name of *Larus glaucus*. In point of priority, therefore, this name ought to be adopted for the present species, in preference to that of *Larus arcticus* given to it by Mr. Macgillivray, or that of *L. leucopterus*, under which it is described by Richardson and Swainson in the Fauna Boreali-Americana, and by the Prince of Musignano in his Synopsis. Captain Sabine in his memoir on the Birds of Greenland, in the twelfth volume of the Linnean Transactions, has described the same bird under the title of *Larus argentatus*, and this in deference to the opinion of M. Temminck, who at that time considered it as a variety of the *Herring Gull*, occasioned by the rigours of a polar climate. The fact, however, of the true *L. argentatus* having been found with its characteristic markings unchanged in those regions, together with the perfect and undeviating whiteness of the wings of the other bird, and the difference of proportions, observable in the bills of the two species, might justly have made the former author hesitate, before yielding even to the authority of a naturalist so deservedly eminent. The present species, in all its states of plumage from adolescence to maturity, bears the closest resemblance to the Glaucous Gull, and can only be distinguished by its striking inferiority of size, and by the greater length of its wings, which reach, when closed, upwards of an inch beyond the end of the tail; whereas in the other bird they scarcely reach that part. Like its prototype it is a winter visitant to the Shetland Isles and the northern parts of Scotland, and a few occasionally stray as far southward as the Northumbrian coast, where I have obtained three or four specimens, but all in the immature plumage. Its habits are stated by Mr. Edmonston to be more lively than those of the Glaucous Gull, and it displays more elegance of form. It is a common species in the arctic regions, and is mentioned by Sabine and Richardson as being plentiful in Baffin's Bay, Davis's Straits, and Melville Island. It is also common on the Iceland coast, to which it is probable many of those that winter with us, and in similar latitudes, retire to breed. It feeds upon fish, the flesh of whales, and other carrion, and when upon our shores is sometimes seen in company with the Black-backed Gull."

In summer the adults have the head, neck, tail, and under surface pure white; the mantle and wing-coverts pale grey; the shafts and tips of the quills pure white; bill pale reddish flesh-colour at the base, and black at the tip; feet pale flesh-colour; irides pale yellowish grey.

In winter the head and neck are streaked with grey.

The young have the entire plumage pale yellowish grey barred and mottled with pale brown; the quills greyish white tinged with brown; and the tail dull brown marbled with white.

We have figured an adult male rather less than the natural size.

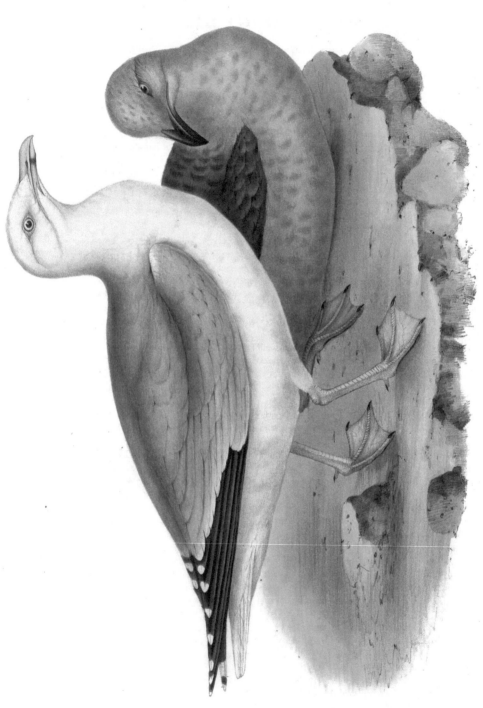

HERRING GULL.
Larus Argentatus. (Brown.)

Printed by C. Hullmandel.

Drawn from Life and on Stone by J.& E. Gould.

HERRING GULL.

Larus argentatus, *Brunn.*

Le Goêland à manteau bleu.

THE Herring Gull is very abundant along the shores of Great Britain, as well as those of the European Continent. It remains with us during the whole of the year, occasionally visiting our lakes, rivers, and inland waters. The British Islands and the coast of Holland may be considered nearly the extent of its range southwards. M. Temminck informs us, that although the young are occasionally found along the shores of the Mediterranean, the adults are very seldom to be seen there. They breed along the rocky parts of our own coast, particularly at the Isle of Wight, from Fresh-water Gate to the pointed rocks called the Needles; the coasts of Wales, Scotland and the adjacent islands; as well as the lengthened and precipitous coast of Norway and the shores of the Baltic generally. At the time of incubation, these birds assemble in numerous companies, often associating with Guillemots, Razor-bills, and Puffins. They compose their nest of marine vegetables, which is placed on the ledges of rocks, or elevations covered with herbage, as samphire, grasses, &c., the female laying two or three eggs, about two inches and a half long, by one inch and three quarters in breadth, of a greenish olive colour spotted with black and ash brown; in the depth of ground colour and disposition of the markings there is, however, great variety. Independently of the variation in plumage which this bird undergoes in passing from youth to the adult state, there is another change which annually takes place in mature birds,—a deviation from the usual law which under similar circumstances is found to occur.

In the winter, the head, neck and chest no longer retain the pure white which forms the livery of the breeding season, but each of the feathers which cover these parts becomes streaked with a longitudinal mark of brown, so as to give it a mottled appearance; the rest of the plumage remains unaltered. The top of the back and scapulars are of a pure blueish ash; the quills black, each feather tipped with white; the rump, tail and whole of the under surface uniform white; beak yellow; the under mandible has the angular projection of a bright red; the naked skin round the eyes yellow; irides delicate straw yellow; legs and feet flesh-coloured. Length about twenty-two inches. It is not before the third year, at least, that the Herring Gull attains its perfect state of plumage. The young at first have the head, neck, and all the under parts grey mottled with light brown; the upper parts light ash-brown; tail-feathers whitish at the base, becoming gradually brown to their termination; quill-feathers blackish brown just tipped with white; beak dark brown or horn-colour; naked circle round the eye and irides brown; feet blueish brown. From this stage they may be seen in all their intermediate degrees, up to that of maturity, which may be considered as perfect in the fourth year;—it is, however, more than probable that before this ultimate change they commence breeding, as we have seen them at nest with the remains of the colours of nonage intermingled with the white and blue of maturity.

Our Plate represents an adult male, and a young bird in the second year, two thirds of their natural size.

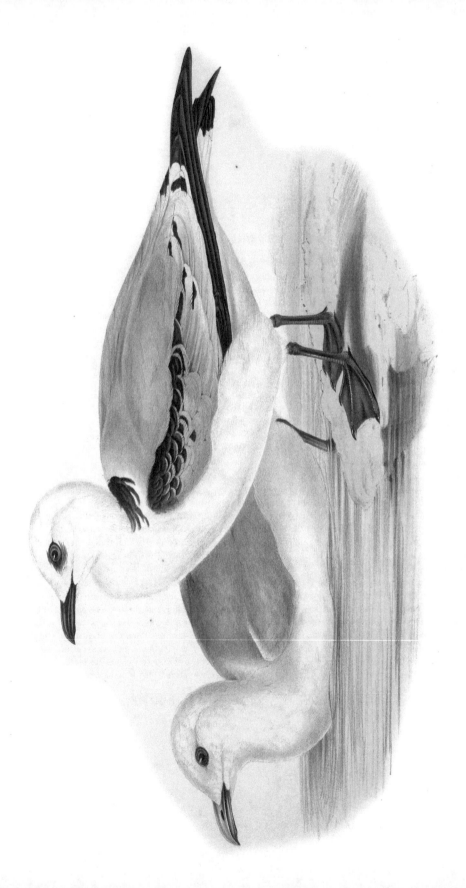

KITTIWAKE GULL.
Larus Rissa. (Linn.)

Drawn from Nature & on Stone by J & E. Gould.

Printed by C. Hullmandel.

KITTIWAKE GULL.

Larus Rissa, *Linn.*

La Mouette tridactyle.

It would appear that we must consider this species of Gull as only one of the list of our summer birds of passage, as it journeys south in the winter, and returns again in the spring to its usual haunts. Whether this is the case with the birds of this species in the continental parts of Europe we are not able to say; but as it is spread far northward along the whole of the European shores, we may conjecture that at least in the higher latitudes it is a bird of migratory habits. We do not consider that the abbreviated hind toe, which is a distinguishing characteristic of this Gull, is of sufficient consequence to entitle it to rank as a distinct genus; and though Mr. Stephens has thought differently, we are not inclined to adopt the term *Rissa* as a generic title, but retain the word as its specific appellation.

The difference which the plumage of the Kittiwake exhibits at different ages has led to a multiplication of its synonyms and some degree of confusion, the young having been considered by many ornithologists as a separate species, and described under the title of *Larus tridactylus*, and in popular language the Tarrock; this error, like others of the same kind, which in the works of the earlier writers were almost unavoidable, is now cleared up, the various gradations of plumage from youth to maturity being well ascertained.

In its habits and manners the Kittiwake generally resembles the rest of its congeners; it is, however, less addicted to seeking its food on the land, but is observed ever busily engaged over the surface of the water, in pursuit of small fishes, mollusca, crustacea, and other aquatic productions, which constitute its means of subsistence.

The places chosen for its sites of incubation are the ledges of bold precipitous rocks overhanging the sea : numbers breed annually on the Farn islands, at Flamborough Head, on the Bass Rock; many also breed annually about Freshwater, Portland Island, and elsewhere. The nest is made of dried grass and sea-weed, and the eggs are two in number, of an olive white, blotched with dark brown and purplish grey.

The common name of Kittiwake is given to this bird from the peculiar call during the season of incubation, which the male reiterates as he wheels round his mate upon the nest, or pursues his way on buoyant wing over the surface of the waves.

In its adult stage, which is not attained till the second autumn, the plumage of the Kittiwake is very simple, the mantle and wing-coverts being fine pearl grey; the quills are tipped and bordered along their outer margin with black; the head, neck, tail, and under surface white; bill yellow; tarsi and toes dark olive green.

The young of the year have the bill black; head, neck, chest, and under parts white, with the exception of a black spot near the eye and nearly encircling it; a marked crescent of black crosses the upper part of the back, and advances upon the neck; the rest of the back and scapulars are grey; the lesser wing-coverts black; the greater coverts and secondaries grey, passing into dull white, with terminal patches of black; tail white, largely tipped with black.

After the first general moult the black markings become more obscure and limited, and the bill acquires a tinge of olive; at the next autumn moult, that is, in twelve months after the first, the full plumage is acquired.

Our Plate represents an adult bird and a young bird of the year, of the natural size.

IVORY GULL.
Larus eburneus. (Gmel.)

Drawn from Nature & on Stone by J & E Gould.

Printed by C. Hullmandel.

IVORY GULL.

Larus eburneus, *Linn.*

La Mouette blanche, ou Sénateur.

FROM the circumstance of two or three examples of this beautiful Gull having been captured at different times within the precincts of the British Islands, all modern writers have included it in the Fauna of this country. The snowy whiteness of its plumage renders it one of the most delicate and interesting species of its genus. Dwelling almost solely within the regions of the arctic circle, the few stragglers which now and then pass the boundary line, and visit the more temperate portions of the European continent, are, if taken by the ornithologist during these peregrinations, considered a prize of no little rarity and value. The first authenticated instance of its being captured in the British Islands was communicated to the Wernerian Society by L. Edmonston, Esq., and a notice of the occurrence published in the fourth volume of the Memoirs of that Society. This individual, which was killed in Balta Sound, Shetland, in December 1822, and one since, in an immature state, in the Frith of Clyde, are the only recorded instances of its having been found near our coasts. In a note in the Manuel of M. Temminck, we find this author also expatiating on its extreme rarity in our latitudes, two individuals only having at that period come under his notice.

From the accounts given of this Gull in the works of Dr. Richardson, Capt. Sabine, and most arctic voyagers, we learn that in those regions it is a species of no rarity; and from its being equally common in Greenland and Spitzbergen, we may naturally conclude that it ranges over the whole of the arctic circle. In these solitary wilds it is constantly accompanied by the Fulmar Petrel; and like the generality of its tribe, which are constantly observed in the neighbourhood of shipping, it is always to be seen following the whalers and feeding upon the refuse thrown overboard, which, with blubber, small fish, and crustacea, forms the principal portion of its diet.

It is said to breed in rocks overhanging the sea, but the number and colour of its eggs we have yet to discover.

The sexes, when fully adult, are alike in colouring; the young, on the contrary, (as is the case with most species of the genus,) are so very dissimilar that they have been mistaken, and described as a distinct species. The plumage of the first autumn is an almost uniform blackish grey, which gradually gives place to a mottled livery of black and white, the ends of the primaries and tail retaining the dark marking the longest, and until the end of the second year. It is said that the immaculate white plumage is that of summer, and that the head and neck are streaked with grey in winter.

The base of the bill is deep lead colour, the remainder being fine ochre yellow; the irides are brown; the feet black; and, as the name implies, the whole of the plumage is pure white.

The Plate represents an adult male rather more than three fourths of the natural size.

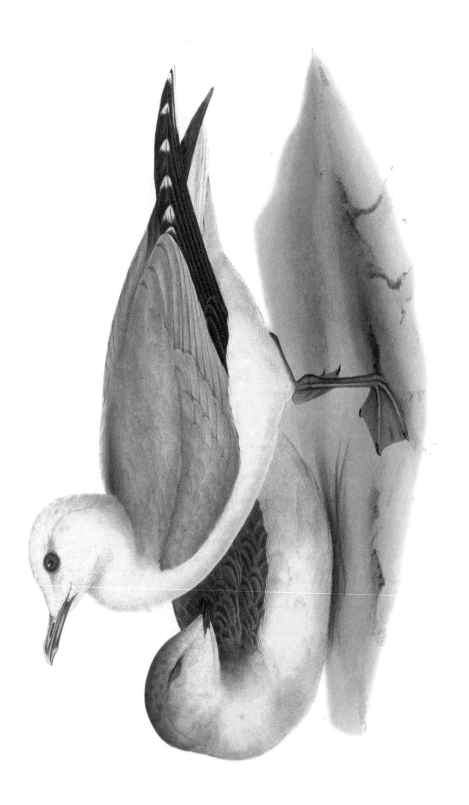

COMMON GULL.

Larus canus: *(Linn.)*

Drawn from Nature & on stone by J & E Gould.

Printed by C Hullmandel.

COMMON GULL.

Larus canus, *Linn.*

La Mouette à pieds bleus.

THE Common Gull, as its name implies, is one of the most abundant species inhabiting the British seas, there being no part of our coast on which it may not be found; it is, moreover, a resident species, breeding, according to Mr. Selby, on bold rocky headlands "overhanging the sea, and sometimes on islands; or on the shores of lakes, as I have found in two or three instances in the Western Highlands of Scotland. At St. Abb's Head, a bold and rocky headland of Berwickshire, these birds are very numerous during the breeding-season, and occupy the whole of the face of the cliff."

The nest is formed of sea-weed and grasses; the eggs, which are two and sometimes three in number, are of a yellowish white, blotched irregularly with brown and grey.

Like many of the other members of the genus, this bird is two, if not three years before it attains its perfect plumage, the change being from mottled greyish brown to a delicate lead colour on the upper surface and white beneath.

It is sometimes seen, particularly in the winter season, at a considerable distance from the shore, and it is often known, like the Rook, to follow the plough, and to wander in small flocks over fallow lands in search of worms, insects, and their larvæ.

It is said to possess an extensive range, and to pass the summer in most of the arctic regions, inhabiting equally those of North America, Europe, and Asia, whence on the approach of winter it migrates southward, and inhabits for a time most of the temperate parts of Europe.

In winter the head, occiput, nape and sides of the neck are white streaked with brown; the mantle, scapularies, and wing-coverts pearl grey; primaries black towards their tips, which are white, and the two first have also a large white spot within the black; under surface, rump, and tail pure white; bill bluish green at the base passing into ochreous yellow towards the point; gape orange red; naked skin round the eyes reddish brown.

In spring the brown streaks on the head and neck disappear, and those parts become of a perfectly pure white; the bill changes to a deeper yellow, and the eyelids to bright vermilion.

As above mentioned, the young are at first mottled with greyish brown, grey and white, which is gradually exchanged at successive moultings to the adult plumage; the legs and toes are pinkish grey; the base of the bill fleshy red, and the tip blackish brown.

We have figured an adult and a young bird rather less than the natural size.

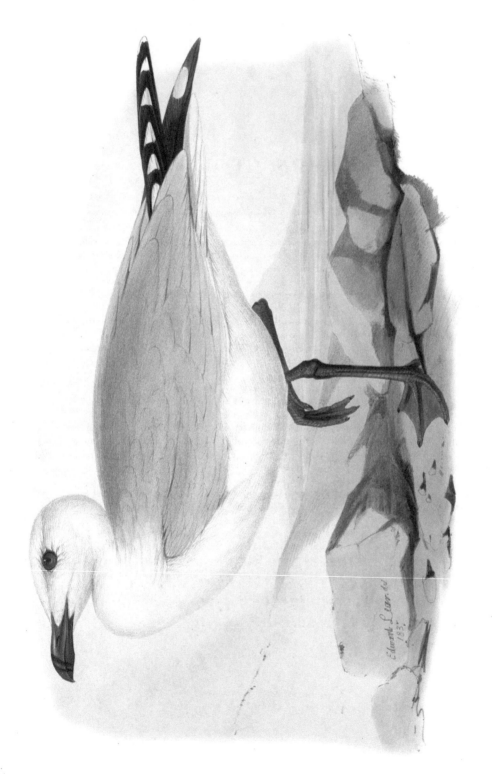

AUDOUIN'S GULL.
Larus Audouini. *(Payr.)*

Edward Lear del
183?

AUDOUIN'S GULL.

Larus Audouinii, *Temm.*

La Mouette d'Audouin.

ALTHOUGH we are not aware of any instance of the occurrence of this fine species of Gull in our seas, still from a letter we have lately received from our friend M. J. Natterer, it would appear that it is by no means rare in the Mediterranean, for says M. Natterer, " I shot three of these gulls near Gibraltar and Tarifa, the whole of which had white heads in the month of August, the species cannot therefore belong to that section of the family which during this month have the head black."

From our knowledge of birds, we should say that the present species is extremely local, and we have never observed it in any of the many foreign collections we have had opportunities of examining. Our figure is taken from a fine specimen sent to us by M. Temminck, but from what locality it was obtained is not stated. It is probable that independently of those of the Mediterranean the whole of the coasts of Northern and Western Africa constitutes its native habitat.

The situation of the nostrils in this species, together with the absence of the black head in summer, sufficiently indicates its separation, as M. Natterer has observed, from those gulls which we have included under the generic title of *Xema*.

Head, neck, all the under surface and tail pure white; mantle and wings pale silvery grey; primaries black, tipped with white; bill and legs red, the former crossed near the tip with two stripes of black.

We have figured a male in the summer plumage nearly of the natural size.

SKUA GULL.

Lestris catarractes. (*Temm.*)

Drawn from Life and on Stone by M.E.Gould.

Printed by C. Hullmandel.

Genus LESTRIS.

GEN. CHAR. *Beak* moderate, hard, strong, cylindrical, very compressed, hooked at the point, the *upper mandible* covered with a cere, the *under mandible* with an angle on the inferior edge. *Nostrils* approaching the point of the beak, diagonal, narrow, closed on their posterior part, and pervious. *Tarsi* long, naked above the knee. *Feet* having three toes before, entirely palmated; hind *toe* very small; *nails* large and hooked. *Tail* slightly rounded, two middle feathers elongated. *Wings*, first *quill-feather* longest.

SKUA.

Lestris catarractes, *Temm.*

Le Stercoraire cataracte.

THE Skua is an inhabitant of the higher regions of both hemispheres: it is constantly found on the Northern seas of the European Continent; and although it is not met with, we believe, in the North American seas, Captain Cook observed it at the extremity of the Southern Continent, being very abundant about the Falkland Islands; and several collected by Captain P. P. King, on his last survey of the Straits of Magellan, and Terra del Fuego, were found on examination to be strictly identical with our own. In Europe; the Orkney, Shetland and Feroe Isles appear to be among the favourite breeding-places, and during the period of incubation the male becomes extremely fierce and pugnacious; it is, notwithstanding, a welcome guest to the inhabitants, whose flocks, but for this bird, would be more frequently exposed to the ravages of the eagle and raven; the former he will courageously attack, and repel, whenever he appears within the range of his dominions, for which service we can personally testify to the unwillingness with which the natives allow this bird to be destroyed.

The *Lestris catarractes* may be often observed wandering about, generally in pairs, on the northern shores of these Islands; the season, however, when the Skua may be most abundantly met with, is that in which the innumerable shoals of herrings visit our shores, at which times they are followed by flocks of Gulls of various species, who find in them an ample repast. It is not, however, for the sake of fishing, himself, that the Skua follows in the train, but, like the rest of his congeners, for the purpose of depriving the more industrious labourers of their booty, harassing them with unceasing ferocity until they deliver up their spoil. Fish thus obtained is not, however, his only food; for carrion, and the flesh of dead cetaceous and molluscous animals are not refused; it is even asserted, that, like the rapacious tribe of Falcons, whose place he may be said to take on the ocean, he will destroy birds of inferior size and strength,—an act for which his formidable talons, strong hooked beak, and great powers of flight, render him extremely well qualified.

The sexes differ but little in colour and size, and, contrary to what takes place in the other species of this genus, the young and adult exhibit but trifling variations. Its entire length is about twenty-two inches.

The bill is long, black, and strongly hooked; legs and feet jet black; length of the tarsi two inches and a half; feet webbed; toes armed with strong hooked talons, that on the inner toe being the most formidable Its general colour is a dark umber brown, varied on the back with light shades of reddish brown; the neck marked with elongated lines of dusky yellow; the first quill-feather of the wings the longest, the upper part of the webs and shafts white; the lower part dark brown; tail cuneiform, the two centre feathers projecting about an inch beyond the others. The egg of this bird resembles that of the Herring Gull in shape and colour, but is rather smaller, measuring two inches and a half in length by one and three quarters in breadth, olive brown blotched and spotted with darker brown.

We have figured an adult male, two thirds its natural size.

In our description of the Pomarine Gull we omitted to notice the egg of that bird, which, as figured by Naumann and Buhle, measures two inches three lines in length, and one inch eight lines in breadth; of a dark olive green, blotched and spotted with two shades of red brown.

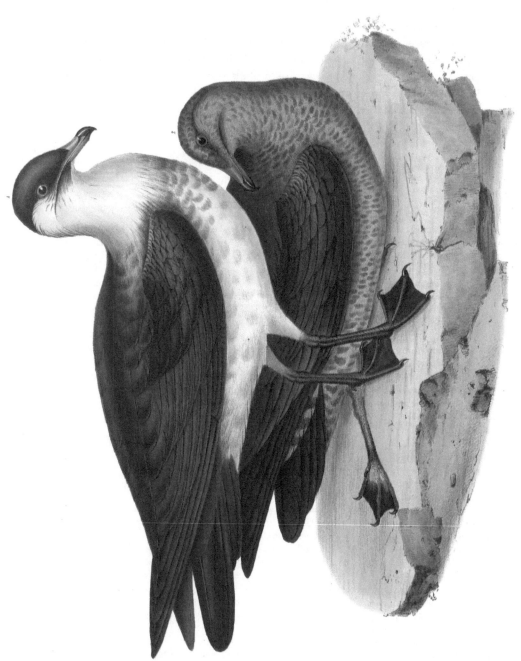

POMERINE GULL.

Lestris pomarinus. — *Temm.*

Drawn from Life on Stone by J.J.Gould

Printed by C.Hullmandel

POMARINE GULL.

Lestris pomarinus, *Temm.*

Stercoraire pomarin.

AMONG the parasitic Gulls, the *Lestris pomarinus* is the second in size, and possesses much the same habits as the preceding species, preferring a life of plunder to one of quiet industry : hence has arisen the generic title of the family, the word *lestris* signifying 'a robber'. We are indebted more particularly to M. Temminck (who appears to have been the first to characterize it,) for our knowledge of the present species, which, from the various changes it undergoes, had previously occasioned no little confusion. It is an inhabitant of the more northern regions of both continents ; but on the European side the rocky and extensive coasts of Norway and Sweden are the only localities where it is supposed to incubate. The adult birds appear to confine themselves to the districts where they build, and are rarely met with on our shores, or those of the southern countries of Europe. The young, however, wander very extensively; abounding at certain seasons on our own seas, and along the coast of France ; and we further learn, on the authority of M. Temminck, that the Rhine, and the lakes of Switzerland and Germany are also visited by them. We have procured these birds in considerable abundance, from every part of our own coast, while in the performance of their extensive migrations, at which time they have so close a resemblance in colour to the well-known blacktoed Gull, as to be easily mistaken for that bird ; from which, however, they may be distinguished by their greater size : the beak in this species is also much more robust ; the tarsi longer and more roughly reticulated. But, before entering more fully into a description of the present species, we must not omit to notice a peculiar characteristic in this class of Gulls, consisting in the length of the two middle tail-feathers, which extend beyond the others. In the *Skua* their length but little exceeds that of the tail, their breadth at the base continuing the same to the end, which is squared ; in the present species the length is increased, the breadth continuing the same, but the end rounded : in the remaining species of this genus the two middle tail-feathers are extensively prolonged, gradually tapering from the base and terminating in a point. The beak of the adult male is of a greenish yellow ending in a black point, which is much curved ; the irides yellowish brown ; the feet and webs deep black ; a blackish brown covers the head, face, and occiput, where it terminates in a point ; throat white ; cheeks and sides of the neck covered by silky filamentous feathers of a delicate straw-colour ; the whole of the upper surface, wings and tail, of a deep umber brown ; the chest thickly clouded with irregular bars of brown, becoming lighter towards the belly, which is white ; vent and under tail-coverts brown interspersed with white ; the middle tail-feathers exceed the rest by two or three inches. The total length of the bird is fifteen or sixteen inches ; the adult male and female resemble each other in plumage.

The young of the year present a uniformity of colouring throughout, which consists of a dark brown, each feather being tipped with ferruginous brown : in this stage, the middle tail-feathers scarcely exceed the others ; as they advance in age, the adult plumage gradually supervenes. It is in the intermediate state that the Pomarine Gull is most commonly to be met with.

Our Plate, in which the figures are two thirds of their natural size, represents an adult male in full plumage, and a young bird of about the age of five months in its immature dress.

RICHARDSON'S LESTRIS.

Lestris Richardsonii. *(Swains.)*

RICHARDSON'S LESTRIS.

Lestris Richardsonii, *Swains.*

THIS is by far the most common parasitic Gull on our coast. It breeds in the Orkney, Shetland, and Western Isles in very considerable abundance ; and from this, its most southern boundary of incubation, it may be found in all the intermediate countries to the polar regions of both continents. It was discovered by Dr. Richardson, breeding on the barren ground north of Hudson's Bay at a considerable distance from the sea, from whence he brought specimens, which we have had opportunities of comparing with others procured in the Orkneys, and find them to be strictly identical. Some confusion appears to have existed in reference to the specific differences of these birds ; and had it been consistent with the plan of the present work, it would have been a gratification to us to have figured all the known species contained in this interesting and well-defined genus : they are, however, not numerous ;—we are not acquainted with more than two others extra-European, both of which are natives of America. The present species has hitherto been considered as the *Larus parasiticus* of Linneus ; and it is somewhat surprising, that its claims to a distinct specific title should so long have escaped the notice of European Ornithologists, and especially the scientific and discriminating eye of M. Temminck. We have had opportunities of examining numerous examples of both species, and also specimens of the bird figured by Edwards under the name of the Arctic Bird, Plate 148, which has legs of a bright yellow colour, and tail-feathers much more elongated. This bird we also consider to differ both from *L. Richardsonii* and *L. parasiticus*; but as it has never been known to visit Europe, further description must, in accordance with our plan, be omitted.

The *Lestris Richardsonii* first received its specific title from Mr. Swainson, in honour of Dr. Richardson ; a figure and description of it being published in the Fauna Boreali-Americana, part 2, p. 453 ; and it is with no small degree of pleasure, that our work affords us an opportunity of illustrating a bird bearing the name of so distinguished a traveller, whose arduous and indefatigable exertions have done so much for science.

The *Lestris Richardsonii* is subject to so great a variety of changes in the colour of its plumage, that a more than usually minute detail in the description is required. The dark chocolate coloured bird figured in our Plate, characterizes at least three fourths of the specimens found breeding in the British Isles ; and from this colour to that of the light-coloured bird on the same Plate, it may be seen in all the intervening shades. Both sexes appear to be subject to the same law, and a very light-coloured male may be often seen paired with a dark female, and *vice versà*. We are therefore led to believe, from the circumstance of the darker-coloured birds forming the greater portion, that this colour prevails solely among young birds, yet sufficiently matured for the reproduction of the species. The young birds of the year present precisely the same disposition of markings and colouring which is so characteristic of the *Lestris Pomarinus*, being brown, numerously barred with transverse lines of a richer colour ; the legs and a portion of that part of the web nearest the tarsus are flesh-coloured ; the other part of the membrane is black : this very conspicuous character formerly obtained for this species the name of Black-toed Gull (*Larus crepidatus*), an appellation which has long sunk into a synonyme. The *Lestris Richardsonii* is a more robust and powerful bird than the *Lestris parasiticus* ; and the upper surface of its plumage is darker and more uniform in colour ; the two middle tail-feathers scarcely ever exceed the others more than three inches ; and the tarsus, toes, and interdigital membrane are also conspicuously larger.

Total length 21 inches ; wing 13 inches ; middle tail-feathers 9 inches long, exceeding the rest of the tail 3 inches ; beak 1¼ inch ; tarsus 1 inch 9 lines.

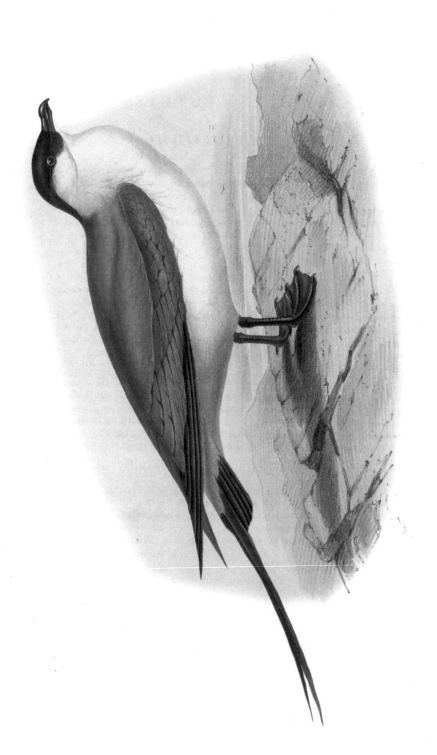

PARASITIC GULL.

Lestris parasiticus. (Ill. Temm.)

Printed by C Hullmandel.

Drawn from nature on stone by J. E. Gould.

PARASITIC GULL.

Lestris parasiticus, *Ill.*

Le Stercoraire parasite.

THIS very elegant and delicately coloured species of *Lestris* is, we believe, the true *parasiticus* of Linneus, Buffon, and Temminck. On comparison it will be found to differ very materially from the *Lestris Richardsonii*, but to which it forms the nearest approach, exhibiting, however, a well-defined specific character. Nothing can be more beautiful and complete than the regular gradation which characterizes the species of the genus *Lestris* : commencing with the Skua, which is the largest, a concatenation may be observed throughout the whole. As the species diminish in size, the elongated tail-feathers are still more lengthened. This character is carried to an extreme in an American species, which is rather smaller than *L. parasiticus*, and has the central tail-feathers at least five inches longer, while in the Skua they scarcely extend beyond the remainder of the tail.

We have not been able to ascertain whether the *Lestris parasiticus* breeds among the British Isles, and it is certainly of rare occurrence. Its natural habitat appears to be more confined to the North, viz., the shores of the Baltic Sea, the rugged coasts of Norway and the polar regions. M. Temminck informs us, that it migrates periodically into Germany, Holland and France, but mostly in its immature state. It feeds on fish, insects, and portions of dead cetacea, which it usually procures by harassing and buffeting unfortunate Gulls and Terns, until they are obliged to lighten their bodies by disgorging half-digested fish, &c., the fruits of the labour and search of several hours.

Although it is probable that this bird undergoes variations in plumage similar to those of the *Lestris Richardsonii*, we are by no means able from our own knowledge to state this to be the case; as in all the specimens which we have had opportunities of examining, the markings have been clear and decided, the birds exhibiting a well-defined dark-coloured cap on the head, light under parts, and very long middle tail-feathers.

The female differs but little in plumage from the male, and the young resemble in colouring the other species of the genus of the same age.

The top of the head and the space between the bill and the eyes of a deep blackish brown, terminating at the occiput, the whole of the upper surface of a clear brownish grey; quill- and tail-feathers much darker; the throat, neck, and under surface of a pure white, with the exception of the cheeks and sides of the neck, which are tinged with a delicate straw-yellow; legs and feet black.

Total length 21 inches; wing 11¾ inches; middle tail-feathers 12 inches, exceeding the rest of the tail 6¾ inches; beak 1⅝ inch; tarsus 1¼ inch.

We have figured an adult male three fourths of the natural size.

MANKS SHEARWATER.
Puffinus Anglorum./*Ray./*

MANKS SHEARWATER.

Puffinus Anglorum, *Ray*.

Le Pétrel Manks.

So exclusively aquatic is this little tenant of the ocean, that the impulse of incubation alone induces it to visit terra firma, on which occasion it resorts to those portions only of the land which are washed by the surge, generally selecting such places as small islands, which, from the danger of approach and their rocky nature, are seldom made the residence of man. Whenever he does take up his abode there, the number of birds speedily diminishes; and this is clearly shown by the total absence at the present period of this species in the Calf of Man, where in the time of Willughby and even at a later date they appear to have been very abundant. Mr. Selby is inclined to believe, and we fully concur in his opinion, that this diminution of their numbers is wholly occasioned by the wanton and greedy destruction of their eggs and young, which are eagerly sought after as an article of food, the latter being considered by many a great delicacy, and eaten both fresh and salted.

It is evident that the author above mentioned is not aware that this highly interesting bird is even now, during the months of summer, nearly as abundant on the coasts of South Wales as it was formerly in the Calf of Man. We are fully borne out in this assertion by the circumstance of our having received from thence, through the medium of a friend, no less than four dozens of these harmless creatures at one time, with an assurance that as many more would be forwarded if required. These were all evidently captured by the hand, none of them possessing any of the usual indications of having been shot. From what information we could obtain, it appears that the Manks Shearwater visits these localities for the purpose of incubation during the early part of spring, when they resort to deserted rabbit-burrows, crevices of the rocks, &c., wherein they deposit their single white egg, and the birds then fall an easy prey to the fishermen and others. Giving a decided preference to the western coasts of our islands, they are tolerably abundant in Ireland and in the Western and Orkney Islands. After the conclusion of the breeding-season they retire southwards, even beyond the Mediterranean, where, in consequence of the increased temperature, they find a greater supply of food than they could in more rigorous climates during the season of winter.

Their food consists of all kinds of marine animal substances, such as crustacea, small fishes, mollusca, &c.

In its general contour, the Manks Shearwater is admirably adapted for traversing the surface of the ocean, and from the lengthened form of its wings it undoubtedly possesses great power of flight. The coast of Norway and the shores of the Baltic, although not without the presence of the Shearwater, appear to be much less frequented by it than our own islands; and in the "Manuel" of M. Temminck it is stated to be a bird of very rare occurrence on the shores of Holland and France.

The sexes are alike in the colour of their plumage, and the young resemble the adults at an early age.

The head and whole of the upper surface is of a dark brownish black; the neck, chin, and throat transversely marked with indistinct lines of the same colour; all the remainder of the plumage white, with the exception of a spot of blackish brown behind the thighs; bill yellowish brown at the base, and dark brown at the tip; legs and feet brown; irides hazel.

The Plate represents an adult male of the natural size.

DUSKY SHEARWATER.
Puffinus obscurus.

Drawn from Nature & on Stone by J.& E.Gould.

Printed by C.Hullmandel.

DUSKY SHEARWATER.

Puffinus obscurus.

Le Petrel obscur.

THE Dusky Shearwater so closely resembles the preceding species both in form and colouring that its diminutive size may be said to constitute the only difference by which it is distinguished from that bird; no doubt, however, exists in our minds as to their being really distinct. The two species are moreover inhabitants of different parts of the globe, the *Puffinus Anglorum* being almost confined to the northern seas, while the *Puffinus obscurus* is equally confined to the southern, and rarely found further north than the Mediterranean, on the European shores of which sea most of the European examples have been procured. It is more abundant on the shores of Africa, extending from the Cape of Good Hope to its northern boundary: Africa then may be considered as its natural habitat.

The sexes do not appear to differ in external appearance, nor are the young of the first year distinguished by any particular plumage.

In habits and manners this species is supposed closely to resemble the *Puffinus Anglorum*, but on these points little or nothing is at present known.

Crown of the head, ear-coverts, all the upper surface, wings and tail sooty black; sides of the face and throat transversely marked with indistinct lines of the same colour; all the remainder of the plumage white; bill lead-colour at the base, becoming black towards the tip; feet olive; external web light olive.

We have figured an adult of the natural size.

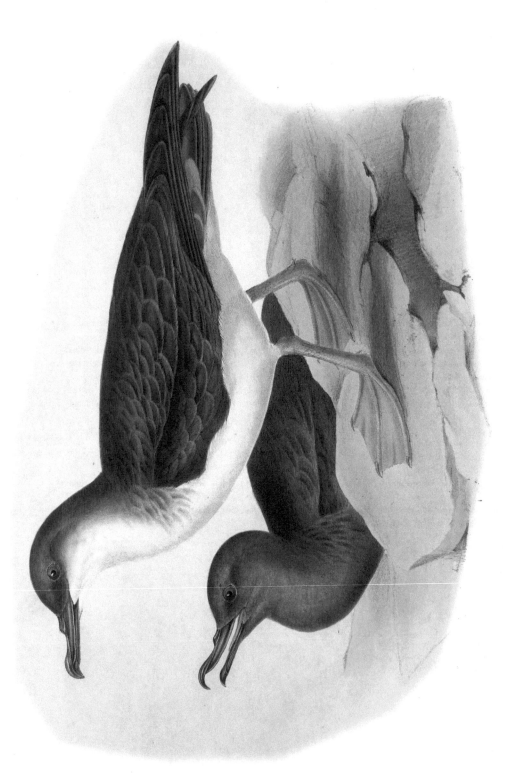

CINEREOUS SHEARWATER.
Puffinus cinereus? (Steph.)

CINEREOUS SHEARWATER.

Puffinus cinereus, *Steph.*

Le Petrel Puffin.

If it should ultimately appear that the bird obtained by Mr. Strickland from the Tees mouth, and charac-terized by him, in the Proceedings of the Zool. Soc. for 1832, under the new specific title of *Puffinus fuliginosus*, is identical with the young of *Puffinus cinereus*, a circumstance which is by no means unlikely; and if a bird apparently in the adult plumage, subsequently obtained by the same gentleman, should prove to be the adult of this species, we shall have, with the addition of a specimen obtained by Mr. Selby, three examples of British-killed specimens of this species. With respect to the specimens forwarded by Mr. Strickland, and which we have figured, we have to observe, that these two birds, although agreeing in their admeasure-ments with each other, differ slightly from a specimen of *Puffinus cinereus* sent to us by M. Temminck as an undoubted example of that species, Mr. Strickland's specimens being less in all their admeasurements; and could we have discovered any difference in the markings of their plumage, we should have had no hesitation in regarding them as distinct: as it is, we have here figured both Mr. Strickland's birds as one and the same species, but with a mark of doubt as to their being examples of the true *Puffinus cinereus*.

The range of the true *Puffinus cinereus* according to M. Temminck is very extensive : " it is spread through-out the Mediterranean, it often appears on the southern coast of Spain and on those of Provence, where many individuals have been killed. It is never seen in the Adriatic. Specimens killed in Senegal and those from the Cape of Good Hope differ in no respect from those killed in Provence. The habits and manners as well as the food of this species do not differ from those of its nearly allied species the Manx Shear-water, *Puffinus Anglorum*, which, as is well known, feeds on all kinds of marine animal matter in a state of putrescency.

The head, cheeks, and all the upper surface pale ash grey, the edges of the feathers on the back being lighter on their external margins, the scapulars, wings, and tail darker; quills deep black; on the sides of the neck and chest are waves of light grey; under-surface pure white; beak yellowish, becoming browner towards the tip; feet and interdigital membrane livid yellow, the webs lightest in colour; irides brown.

Our figures are rather less than the natural size.

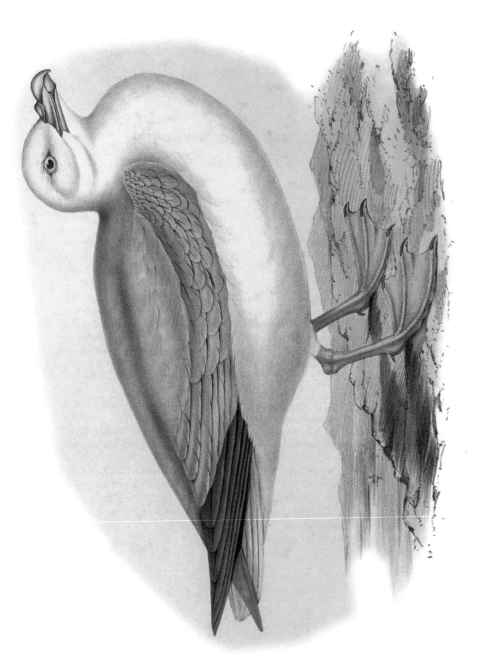

FULMAR PETREL.
Procellaria glacialis. (*Linn.*)

Drawn from life and on Stone by J. & E. Gould.

Printed by C. Hullmandel.

Genus PROCELLARIA.

GEN. CHAR. *Beak* thick, dilated at the tip, sulcated ; the *upper mandible* hooked ; the *lower* straight and slightly truncated. *Nostrils* united in a single tube. *Legs* moderate ; a claw only in place of the hind toe.

FULMAR PETREL.

Procellaria glacialis, *Linn.*

Le Petrel Fulmar.

THE genus *Procellaria*, in which Linnæus placed all the oceanic birds possessing tubular nostrils, and which now form an extensive and well-defined family, has been subsequently divided by naturalists into several minor groups ; and we find that in Europe alone there exist examples which illustrate three genera, viz. *Procellaria* (as now restricted), *Puffinus*, and *Thalassidroma*. The Fulmar Petrel constitutes the type of the genus to which it is assigned, and appears to form the passage to the true Gulls by the interposition of the birds composing the genus *Lestris*. How plainly does the present bird exemplify the wisdom which Nature has exhibited in the creation of all her subjects ! It is in the almost impenetrable polar regions, among floating fields and bergs of ice, often at a great distance from the land, that the Fulmar finds its true and natural habitat ; and in order to enable it to endure the severities of the hardest seasons in these northern latitudes, Nature has afforded it every necessary protection by clothing it in a thick and warm mass of down and feathers of an oily nature, thus precluding cold and moisture. Although the polar regions constitute its native locality, it is nevertheless found, but in much less abundance, in more temperate climates, such as the northern seas of Europe and America, extending itself throughout the lengthened coast of Norway, and not unfrequently Holland and France. It frequents also the northern isles of Great Britain, resorting to the Orkney and Hebrides for the purpose of breeding, but particularly to the Island of St. Kilda.

The food of the Fulmar consists of fish, mollusca, vermes, and the fat of dead cetacea ; it will also devour any oily substance or refuse thrown from vessels, which it fearlessly follows, particularly those engaged in the whale fisheries ; and hence during the season it obtains an easy and bountiful subsistence. They are very active and buoyant on the water, and their powers of flight are considerable.

Their mode of living renders the flesh very disagreeable and unfit for use. Their stomach and body appear to be continually saturated with oil ; and the circumstance of their being able to eject or discharge a quantity of this fluid from their nostrils, when irritated or attacked, is both singular and curious. This power appears to have been given them as a mode of defence, and is characteristic of the whole of the family, from the largest species to the elegant Stormy Petrel ; and even this little creature has the power of squirting out an oily fluid from its nostrils with considerable force.

The Fulmar lays one white egg on the grassy ledges of the rocks and cliffs of our northern islands. They make no nest ; the egg is very large compared with the size of the bird, and has a strong musky smell, which it retains for some time. Our figure represents the plumage of an adult bird ; young birds of the year have the back and wings varied with light grey and brown.

FORK-TAILED STORM-PETREL.
Thalafsidroma Leachii.

COMMON STORM-PETREL.
Thalafsidroma pelagica. (Selby).

Drawn from Nature & on Stone by J & E Gould.

Printed by C Hullmandel.

Genus THALASSIDROMA.

GEN. CHAR. *Bill* shorter than the head; much compressed in front of the nasal sheath, with the tip of the upper mandible suddenly curving and hooking downwards, and that of the lower one slightly angulated and following the curve of the upper. *Nostrils* contained in one tube or sheath, but showing two distinct orifices in front. *Wings* long and acuminate, with the first quill shorter than the third, the second being the longest. *Tail* square or slightly forked. *Legs* having the tarsi rather long and slender, reticulated. *Feet* of three toes, united by a membrane; hind toe represented by a small, straight, dependent nail.

FORK-TAILED STORM PETREL.

Thalassidroma Leachii.

Le Petrel de Leach.

THE first discovery of this Petrel in Europe is due to the researches of Mr. Bullock, who, in the year 1818, while on a tour through the northern and western isles of Scotland, found it breeding on the island of St. Kilda, whence he brought the original specimen from which M. Temminck took his description. In his 'Manuel d'Ornithologie,' under the article alluded to, he dedicates the bird to Dr. Leach by the title of *Procellaria Leachii*. Since the period of its first discovery, when it appeared to be a bird of extreme rarity, it has been found, and that not unfrequently, on most parts of the British coast, and in the channel intervening between our island and the Continent.

In its habits, manners, food, and nidification it so strictly resembles the Common Storm Petrel that the same description will serve for both. It differs from that bird, as also from all other European Storm Petrels, in being one of the largest in size, in having a forked tail, and remarkably short tarsi.

The colour of the plumage is a sooty black, with the under tail-coverts and a patch on the rump white.

COMMON STORM PETREL.

Thalassidroma pelagica, *Selby.*

Le Petrel tempête.

THIS, the least of web-footed birds, though by no means the least important, has been long celebrated by the name of "Mother Carey's Chicken," bestowed upon it by the British sailors, as the foreboder of storm and tempest to the mariner.

The habits and manners of this singular group of birds may be described as being both nocturnal and oceanic. During the bright glare of day they conceal themselves in the crevices of rocks, stones, &c., from whence they depart on the approach of evening, and skim over the surface of the sea in search of food: approaching storms and dull murky weather also rouse them from their retreats to visit their congenial element; hence it is that when seen at a distance from the shore they intimate the approach of gales and severe weather. When out at sea they appear partial to the company of ships, which they follow for days together, and, surprising to say, are never seen to settle on the water; in fact, the only period of rest they appear to allow their organs of flight is while, with extended wing, they skim, half flying half tripping, over the surface of the billows; and it would appear as if the bones of the legs were expressly formed for this manœuvre, being sufficiently flexible to bend without breaking to any opposing pressure or sudden concussion. While skimming around ships they pick up any refuse oily matters that may be thrown overboard, and also any of the small mollusca that may be brought to the surface by the agitation of the water which the vessel occasions in her progress.

The Common Storm Petrel is abundant over the whole of the northern seas of Europe, especially in the rocky islands of Scotland, where it breeds in the crevices of rocks, among loose stones, and occasionally in holes on the ground, generally laying a single egg of a pure white. The young remain in their retreats until their pinions are sufficiently strong for flight, and it is a considerable period before they are able to follow their parents.

The adults of both sexes are alike in plumage, which is invariably of a sooty black, with a white spot on the rump; the bill and tarsi black.

We have figured an adult of each species, of the natural size.

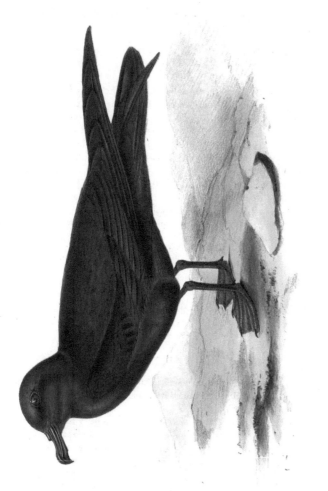

BULWER'S PETREL.
Thalafsidroma? Bulwerii.

Drawn from Nature & on Stone by J & E Gould.

Printed by C. Hullmandel.

BULWER'S PETREL.

Thalassidroma? Bulwerii.

Procellaria Bulwerii, *Jard. & Selby*.

ON the authority of Col. Dalton of Slenningford, near Ripon, we are enabled to add this rare species to the Fauna of Britain, from a fine specimen which was found on the banks of the Ure, near Tanfield in Yorkshire, on the 8th of May, 1837, and which could not have been long dead, as it admitted of being mounted into a good cabinet specimen. It is now in the possession of Col. Dalton, who doubtless regards it as one of the greatest treasures in British ornithology. In fact, with the exception of one or two foreign examples, we do not recollect that we have observed it among the numerous collections we have had opportunities of examining. The only recorded facts relative to its history will be found in the second volume of ' Illustrations of Ornithology,' by Sir William Jardine and Mr. Selby, in which publication the bird is figured from a specimen sent from Madeira by Mr. Bulwer, after whom it has been named. As it is stated to be an inhabitant of Madeira and the adjacent islands, we may infer that the seas bounding the western shores of Africa constitute its true habitat. The cuneated form of its tail and its large size will readily distinguish it from all the other species of its genus.

We cannot conclude our account of this bird without offering our sincere thanks to Colonel Dalton for the loan of his specimen for the purpose of illustration as well as to C. C. Oxley, Esq., of Ripon, who in this instance and upon all other occasions has taken a kind and friendly interest in the present work.

The whole of the plumage is of a deep sooty black, becoming paler upon the throat, and brown on the edges of the greater wing-coverts; bill black; legs and feet blackish brown.

Our figure is of the natural size.

ILLUSTRATIONS

OF

THE FAMILY OF PSITTACIDÆ,

OR

PARROTS:

THE GREATER PART OF THEM

SPECIES HITHERTO UNFIGURED,

CONTAINING

FORTY-TWO LITHOGRAPHIC PLATES,

DRAWN FROM LIFE, AND ON STONE,

BY EDWARD LEAR, A.L.S.

LONDON:

PUBLISHED BY E. LEAR, 61 ALBANY STREET, REGENT'S PARK.

1832.

TO THE

Queen's Most Excellent Majesty,

THIS WORK

IS, BY HER MOST GRACIOUS PERMISSION,

HUMBLY DEDICATED,

BY

HER MAJESTY'S

MOST DUTIFUL AND FAITHFUL SUBJECT,
2915

E. LEAR.

LIST OF PLATES.

PSITTACUS BADICEPS.

Bay-headed Parrot.

E. Lear del. et lith.   Printed by C. Hullmandel.

PLYCTOLOPHUS ROSACEUS.

Salmon-crested Cockatoo. ⅔ Nat. Size

G. Lear del et lith.

Printed by C. Hullmandel.

PLYCTOLOPHUS GALERITUS.

Greater Sulphur-crested Cockatoo.

PLYCTOLOPHUS SULPHUREUS.

Lesser Sulphur-crested Cockatoo.

E. Lear, del et lith.

Printed by C. Hullmandel.

PLYCTOLOPHUS LEADBEATERI.

Leadbeater's Cockatoo.

E. Lear del et lith.

Printed by C. Hullmandel.

CALYPTORHYNCHUS BAUDINII.

Baudin's Cockatoo.

⅔ Nat Size

In the possession of M. Leadbeater

E. Lear del et lith.

Printed by C. Hullmandel

MACROCERCUS ARACANGA.

Red and Yellow Maccaw.

2/3 Nat Size.

E. Lear del. et lith.   Printed by C. Hullmandel.

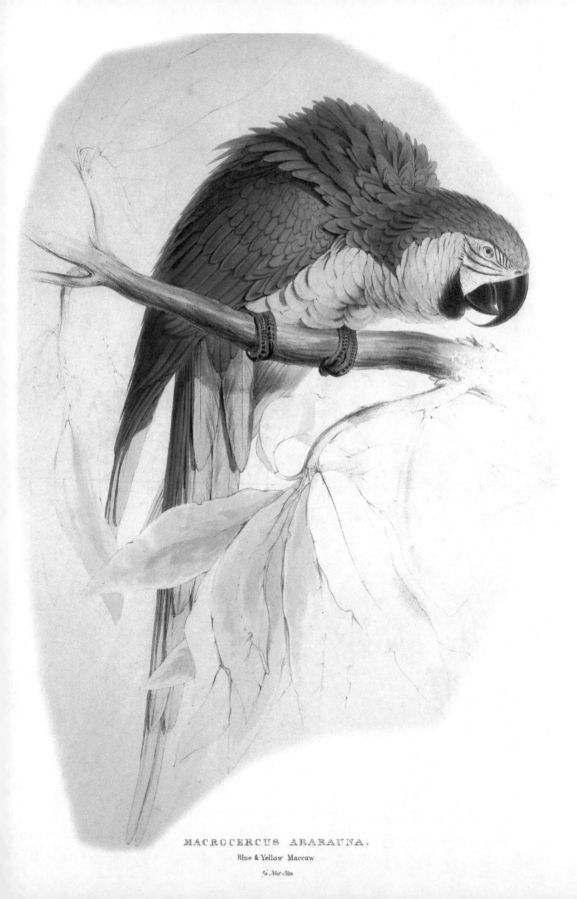

MACROCERCUS ARARAUNA.

Blue & Yellow Maccaw.

½ Nat. Size.

MACROCERCUS HYACINTHINUS.

Hyacinthine Macaw.

E. Lear del. et lith.   3/4 Nat Size.   Printed by C. Hullmandel.

PSITTACARA PATAGONICA.

Patagonian Parrakeet-Maccaw.

E.Lear del. et lith.   Printed by C.Hullmandel.

PSITTACARA LEPTORHYNCHA.

Long-billed Parrakeet-Maccaw.

E. Lear del et lithog.

Printed by C. Hullmandel.

PSITTACARA NANA.

Dwarf Parrakeet Maccaw.

E. Lear del et lith.   Printed by C Hullmandel

NANODES UNDULATUS.

Undulated Parrakeet.

E Lear del et lith.   Printed by C Hullmandel.

PLATYCERCUS ERYTHROPTERUS.

Crimson-winged Parrakeet.

Male Adult.

E. Lear del. et lithog.   Printed by C. Hullmandel.

PLATYCERCUS ERYTHROPTERUS.

Crimson winged Parrakeet.

1. *Female.*   2. *Young Male.*

E. Lear del et lith.

PLATYCERCUS TABUENSIS.

Tabuan Parrakeet.

Printed by C. Hullmandel.

PLATYCERCUS BAUERI.

Bauer's Parrakeet.

E.Lear del et lith.

Printed by C.Hullmandel.

PLATYCERCUS BARNARDI.

Barnard's Parrakeet.

E Lear del et lith.

Printed by C Hulmandel.

PLATYCERCUS PALLICEPS.

Paleheaded Parrakeet

In the possession of Mr Leadbeater

Drawn from nature

Printed by C. Hullmandel

PLATYCERCUS BROWNII.

Brown's Parrakeet.

E. Lear del. et lith.   Printed by C. Hullmandel.

PLATYCERCUS PILEATUS.

Red-capped Parrakeet.

E. Lear del. et lithog.

Printed by C. Hullmandel.

PLATYCERCUS PILEATUS.

Red-capped Parrakeet.

Female.

In the Possession of the Right Hon: Lord Stanley.

E.Lear del et lith.   Printed by C.Hullmandel.

PLATYCERCUS STANLEYII.

Stanley Parrakeet.

E.Lear del et lith.

Printed by C.Hullmandel.

PLATYCERCUS STANLEYII.

Stanley Parrakeet.

Young Male.

E. Lear del et lith.   Printed by C.Hullmandel.

PLATYCERCUS UNICOLOR.

Uniform Parrakeet.

E. Lear del et lith.   Printed by C. Hullmandel.

PLATYCERCUS PACIFICUS.

Pacific Parrakeet.

E. Lear del. et lith.   Printed by C. Hullmandel.

PALÆORNIS NOVÆ-HOLLANDIÆ.

New Holland Parrakeet,

in the Possession of the Right Hon. the Countess of Mountcharles.

1 Male 2 Female.

E. Lear del et lith.   Printed by C. Hullmandel.

PALÆORNIS MELANURA.

Black-tailed Parrakeet.

In the Possession of M.ͬ Leadbeater

E. Lear del et lith.

Printed by C.Hullmandel.

PALÆORNIS ANTHOPEPLUS.

Blossom-feathered Parrakeet.

E. Lear del et lithog.

Printed by C. Hullmandel.

PALÆORNIS ROSACEUS.

Roseate Parrakeet.

E. Lear del et lith.   Printed by C. Hullmandel.

PALÆORNIS COLUMBOIDES.

Pigeon Parrakeet.

E. Lear del. et lithog. Printed by C. Hullmandel.

PALÆORNIS CUCULLATUS.

Hooded Parrakeet.

E. Lear del. et lith.   Printed by C. Hullmandel.

PALÆORNIS TORQUATUS.

RoseRinged Parrakeet._Yellow Variety.

E.Lear del et lith.   Printed by C.Hullmandel.

TRICHOGLOSSUS RUBRITORQUIS.

Scarlet-collared Parrakeet.

E. Lear del. et lithog.

Printed by C. Hullmandel.

TRICHOGLOSSUS MATONI.

Maton's Parrakeet.

E. Lear del et lith.

Printed by C. Hullmandel.

TRICHOGLOSSUS VERSICOLOR.

Variegated Parrakeet.

E. Lear del et lith.

Printed by C. Hullmandel.

LORIUS DOMICELLA.

Black-capped Lory.

E. Lear del et lithog.

Printed by C. Hullmandel.

PSITTACULA KUHLII.

Kuhl's Parrakeet.

E. Lear del et lith.   Printed by C. Hullmandel.

PSITTACULA TARANTA.

Abyssinian Parrakeet.

J. Lear del et lith.

Printed by C. Hullmandel.

PSITTACULA TORQUATA.

Collared Parrakeet.

E. Lear del et lithog.   Printed by C. Hullmandel.

PSITTACULA RUBRIFRONS.

Red fronted Parrakeet.

E. Lear del et lithog.

Printed by C. Hullmandel.

PSITTACULA SWINDERNIANA.

Swindern's Parrakeet.

Edward del et lithog.   Printed by C Hullmandel.